The Pest War

W. W. Fletcher

THE PEST WAR

A HALSTED PRESS BOOK

JOHN WILEY & SONS
NEW YORK

© *Basil Blackwell 1974*

Published in the U.S.A.
by Halsted Press, a Division
of John Wiley & Sons, Inc.
New York

Library of Congress Cataloging in Publication Data

Fletcher, William Whigham, 1918–
The pest war.

"A Halsted Press book."
Bibliography: p.
1. Pest control   2. Pesticides.   I. Title.
SB950. F55   632'.9   74-11440
ISBN 0-470-26401-2

Set in Plantin (text) and Gill Sans (display)
and printed in Great Britain

Contents

Plates

FOREWORD

Many people have made this book possible, but I should like to single out particularly Mr. Arthur Billitt who is well-known in so many spheres of agriculture and horticulture. It was he who first suggested to me that there was a need for a book which, avoiding most of the technicalities of the subject, would nevertheless be authoritative and would have an appeal for the general reader who wished to be informed on this subject. Whether I have succeeded in this objective is not for me to say but Arthur Billitt has been tireless in his encouragement to me and in his hard work on behalf of the book. A number of colleagues and friends have undertaken a critical reading of parts of the book, and I want to express my thanks in particular to Professor W.M. Hutchison, Dr. A. Scott, Dr. R.C. Kirkwood and Dr. M. Burge, all of the Biology Department of the University of Strathclyde; to Mr. W. Newbold and Dr. David Martin of The West of Scotland Agricultural College, Ayr; to Dr. Hubert Martin and to Dr. D. Woodcock of the University of Bristol who guided me through the complexities of chemical nomenclature; to Mr. R.F. Norman, CIBA-Geigy, Dr. H.C. Gough, Ministry of Agriculture, Fisheries and Food, and Professor L. Broadbent, University of Bath, for many helpful criticisms and suggestions; and to Dr. R.A.E. Galley for his invaluable help with Chapter 3. Where possible all sources of information have been included in the Further Reading section but in a book of this kind it is possible that some may have been omitted. If this is so I make my sincere apologies in advance.

A number of appendices have been included, and it is hoped that these will be of value to those wishing to pinpoint particular species of pests or types of chemicals mentioned in the text. Chemical names are from the *Pesticide Manual*, 3rd Edition 1972, ed. H. Martin.

In some of the chapters monetary values may be quoted. In these days of world-wide inflation it is realized that these may be quickly out of date. They should therefore be related to the time in which the book was written and, if necessary, adjusted accordingly.

I am most grateful to the following, who have supplied photographs: the Australian News and Information Bureau (no. 12), CIBA-Geigy (UK) Ltd (nos. 2, 3 and 4), Fisons Ltd (Agrochemical Division) (nos. 7 and 9), May and Baker Ltd (nos. 1 and 11), Mr. Duncan McArthur, Cuan Ferry (no. 10) and Shell Research Ltd (nos. 5, 6 and 8).

Finally, I should like to express my deep thanks to my secretary, Mrs. Isobel Robertson and her assistant, Mrs. E. McPhail, who have laboured long on my behalf, putting up cheerfully with changes and corrections that sometimes entailed the scrapping of a whole day's typing. That the book has appeared at all is due to their hard work.

The whole responsibility for the book rests of course with me. As I have mentioned I have had help, but the ultimate decision as to what went in and what was omitted is mine.

W.W. FLETCHER

INTRODUCTION

The war against pests is a continuing one that man must fight to ensure his survival. Pests (in particular insects) are our major competitors on earth and for the hundreds of thousands of years of our existence they have kept our numbers low and on occasions they have threatened extinction. Throughout the ages man has lived at a bare subsistence level because of the onslaught of pests and the diseases they carry. It is only in comparatively recent times that this picture has begun to alter as, in certain parts of the world, we have gradually gained the upper hand over pests.

This war story describes some of the battles that have been fought and the continuing guerrilla warfare; the type of enemies we are facing and some of their manoeuvres for survival; the weapons that we have at our command ranging from the rather crude ones of the 'bow and arrow' age of pest control to the sophisticated weapons of the present day, including a look into the future of some 'secret weapons' that are in the trial stages; the gains that have been made; and some of the devastation which is a concomitant of war. As with all accounts of war there will be differences of opinion on the interpretation of results and situations; on the emphasis that should be laid on certain parts. This book does not claim to be a definitive account of pest control. It is written for the intelligent non-specialist in the field, who wants to know what the war is all about and the implications of it. It is also aimed at intending students of agriculture, horticulture, medicine, biological sciences; and should also serve as a useful introductory text for elementary courses in pest control in colleges and universities.

Introduction

Although it is non-specialist there has been no attempt to oversimplify. The appendices will be found useful to those wishing to pinpoint the scientific names of pests and pesticides. There is also a bibliography for each chapter. If the books listed prove as useful to the reader as they have been to this writer then the bibliography will have served its purpose.

To Elizabeth

Chapter 1

PESTS

The survival of man through the ages has involved him in a struggle with his physical and biological environment. Subject at first to its vagaries, he has come in time to exercise a partial control over it—over the physical by building more and more sophisticated shelters (from caves to centrally heated apartments) and by the development of suitable clothing; over the biological by improved methods of food production and storage and by combating the pests and disease-causing organisms that threaten his existence. It is a continuing struggle and will be so as long as man is on earth.

This book is concerned with the war that he wages against the pests that threaten his health, and his food supplies. A pest can be defined as any living organism that threatens man's well-being— even fellow men—but in this book we have confined ourselves to four groups: (1) insects and their relatives, (2) weeds, (3) fungi, and (4) certain vertebrates.

In recent times the study of pests has been narrowed to exclude micro-organisms such as viruses, bacteria and fungi which are responsible for causing disease in man, though interestingly enough this does not leave out those insects that may carry many of these disease organisms, nor indeed most of the micro-organisms which attack man's plants.

In their activities pests range from the trivial—a weed on the garden path, a mite causing a slight irritation of the skin—to the disastrous—anopheline mosquitoes carrying malaria, blight fungus destroying fields of potatoes almost overnight. A pest may be either trivial or catastrophic depending upon where it occurs and upon its numbers. The prickly pear for example is of little importance as a pest in its native South America but as we shall see (Chapter 9) it swept through millions of acres of growing land when it was imported to Australia; the migratory locust can be insignificant for many years but may suddenly multiply explosively to devastate large areas of vegetation.

1. INSECT PESTS

There are about one million species of insects—far more than all other animal and plant species combined. The number of individual insects alive at any one time has been estimated at around 10^{18}. Of this vast array 99·9 per cent are from the human point of view quite harmless and some are helpful; the pests are the other 0·1 per cent. Many compete with man for foodstuffs, others carry organisms capable of causing human and animal diseases.

Adult insects are easily recognized in that they usually have the body, which is enveloped in a horny substance called chitin, divided into three distinct regions—head, thorax and abdomen. This separation has given the class its name (Latin *insectus*—cut into). The adult usually has two pairs of wings attached to the second and third segment of the thorax although the true flies have, instead of a second pair of wings, small outgrowths called halteres which are used for balancing. Beetles on the other hand have the first pair represented by wing cases or elytra which protect the delicate flying wings. All insects have three pairs of legs attached to the thorax. They breathe by means of air tubes or tracheae and these air-conducting tubes are distributed like a net all over the body, opening to the outer air by means of paired apertures called spiracles. There are usually two pairs of spiracles in the thorax and ten on the abdomen but there is some variation of number in different species. Insects are often protected by hairs.

The mouths of most insects have one pair of jaws by means of which food is chewed. The other mouthparts are largely sensory and are used to hold or manipulate food prior to action by the jaws. Two outgrowths from the head called feelers or antennae are jointed, usually thin and long and have many nerve endings. They probably serve as touch organs by which impressions are conveyed from one insect to another and they can probably also detect scents. Most adult insects have two kinds of eyes. There are two compound eyes, one on each side of the head, which are made up of numerous six-sided lenses. Between these eyes there may be single-lensed eyes called ocelli which are often disposed in groups of three. Indeed, the insects are well equipped with sense organs, for not only do they have antennae, ocelli and compound eyes but they hear by means of sensory hairs or organs which lie in various parts of the body surface, as well as by tympanal organs (eardrums); and they greatly surpass human beings in their hearing faculties. Insects are also very sensitive to fragrance and to colour and it seems that it is largely by smell that they recognize friends and foes. They may also have a sixth sense which is termed dermatopic—their skins seem to be able to appreciate slight differences of light and shade.

After mating the female stores male sperm inside a special pouch called the spermatheca, and she can use this sperm as required when the eggs are being laid. In some cases the sperm may be stored for as long as five years and still remain capable of fertilizing eggs. Most insects reproduce by laying eggs, the usual number being 100–200 per female. There are extremes, however, some female insects laying only one egg whereas the queen honeybee at her most productive period may lay more than 2,000 per day and during her five-year period of fertility the total may amount to more than one million eggs.

Normally a single individual hatches from one egg, but some of the parasitic insects have developed an interesting mechanism for increasing their numbers. Divisions take place within the egg so that by the time it hatches an egg may give rise to as many as one hundred individuals. This is known as polyembryony.

There are a great many interesting hatching devices, e.g. in the eggs of the bugs there is a cap which is popped open as the individual emerges. Young grasshoppers within their eggs each have a pair of glandular structures called pleuropodia which exude a substance that dissolves the egg shell away. Many insects have a special little spine, the hatching spine on their heads, and this is used to cut their way out of the egg. Butterflies and moths simply chew their way out. Many young increase their internal pressure by swallowing surrounding fluids and air which either causes their heads to swell or causes special little pockets to be erected from their heads. In either case the egg is burst open by pressure. After hatching, insects grow by going through a series of moults, successively casting off their outer skin or cuticle.

3

There are two major groups of insects, classified according to the changes that take place after hatching. In what is considered the more primitive group the newly hatched insect is similar to the adult, except that the wings and the reproductive systems are underdeveloped. These immature stages are known as nymphs and they increase in size and maturity by a series of moults until they become adults with fully developed wings and genitalia. Included in this group are thrips, capsids, grasshoppers, termites, chewing and sucking lice, the true bugs and the aphids. Great damage can be done by some aphids. They bear young in enormous numbers and these young begin feeding from birth. Numerous generations can be produced in any one season. Apart from feeding on plant produce they also render it unsaleable by their excreta, cast-off skins and honeydew. This latter is made up of plant-sap which passes through the aphid, is excreted and forms a supply of foodstuff on which fungi and other organisms which produce disease in plants can multiply. Apart, however, from the direct damage that they do aphids also transmit many viruses from plant to plant: raspberry mosaic, sugar beet yellows, and many potato viruses are among those that are spread within the crop by winged aphids.

The second group of insects is regarded as the more advanced. The individuals go through various stages of development, and some of these stages bear little resemblance to each other or to the final adult insect. The egg hatches to give rise to a larva which feeds voraciously; most damage to plants results from some larva or other feeding on them. Among the most damaging of larvae are the caterpillars of moths and butterflies, the Colorado beetle larvae, wireworms, grain weevil larvae, cabbage-root fly larvae, and leatherjackets which are the larvae of the crane fly.

When the larva is fully grown, the pupa is formed.

Generally the pupa does not feed but fundamental physiological and morphological changes take place within before the adult insect emerges. Many adult insects can also be destructive, among them beetles and weevils which feed on a wide variety of plants, Colorado beetles, wasps and mosquitoes.

Although they are not insects, mites, ticks, millipedes, snails, slugs and eelworms are also generally dealt with by pest control workers.

Mites and Ticks

Both mites and ticks belong to the order *Acarina*. Mites are the smallest members of the order, the largest being the blood-sucking ticks. Unlike true insects they have four pairs of many-jointed legs with claws at the tips. The total number of mites is not known, but there may be in excess of a million species, and they are ubiquitous. The harvest mite is well known. Its larvae penetrate the skin of warm-blooded animals, including man, causing quite a severe irritation. Another mite that attacks humans is the itch mite which burrows in the skin and lays its eggs in the burrows. Some mites transmit the virus of scrub typhus. Economically, probably the most important mite is the red spider mite, species of which attack fruit trees and other crops. In heavy infestations, some may weave fine sheets on the leaves and live within the sheets, sucking the plant juices. Ticks are large blood-sucking mites, and both the larvae and the nymph stages gorge themselves on their victim's blood. Many warm-blooded animals including man are attacked by ticks. Cattle may suffer from tick fever or 'red water', a virulent and often fatal disease transmitted by the tick. In North America, Texas fever, which may cause losses amounting to millions of dollars a year, is passed to cattle by the cattle tick. The Rocky Mountain spotted fever is carried by a mite that infests rabbits, cattle and even man.

Millipedes

Millipedes range from being soft-bodied and less than a tenth of an inch long to heavily armoured forms exceeding eight inches. Many are poisonous. They have two pairs of legs on most of their body segments (centipedes have one pair per segment). Some species have no eyes, but the whole body surface may be light-sensitive. They feed on vegetable matter and some attack crops. Some are luminous. One species, sometimes conspicuous in the redwood forests of California, shines continuously from the time of hatching; the light may be a warning to prospective attackers. Some millipedes may occur in enormous numbers. In 1878, a train was brought to a halt in Hungary by a black mass of millipedes, which carpeted the track and made the wheels slip. Trains were stopped in the same way in France in 1900. In 1918, 75 acres of a West Virginia farmland were covered by millipedes: cattle would not graze and men hoeing the fields became nauseated and dizzy from the smell.

Snails and Slugs

The garden snail has a whorled shell made of chalk and glazed with a protein which tends to wear off with age. Each snail is hermaphroditic, having both male and female organs, but they do mate and lay eggs that hatch in 2–4 weeks. The snail moves by waves of muscular activity passing forwards along the sole of the foot, giving out a slime just behind the head to make its slime track. It hibernates in its shell from September to April and awakes hungry, feeding on the leaves of many plants. The giant snail, which may be up to eight inches long and weigh half a pound, is a pest in many parts of the world—Mauritius, the Seychelles, India, Malaya, Indonesia, the Philippines and China. It feeds voraciously on the leaves, fruit, bark and flowers of a great variety of crop plants—beans, breadfruit, citrus trees, melons, yams and rubber. Needing calcium for its shell, it may climb the walls of houses to attack the whitewash for its lime content. The dwarf pond snail is host to the liver fluke which has in its time caused the deaths of millions of sheep, with an annual loss of millions of pounds sterling. There are also intestinal flukes, lung flukes and three species of blood fluke responsible for the disease *bilharzia* (schistosomiasis), all passing part of their life cycles within snails.

A slug is simply a snail without a shell (though some may have a small shell visible on the surface towards the rear). Many slugs are omnivorous and will eat fungi, carrion, dung, kitchen refuse, worms, millipedes and plants (including crop plants such as vegetables).

Eelworms

Eelworms are microscopic in size, most of them being only about a millimetre in length. They are found in a variety of places and some are parasitic, particularly on plants. Most are very similar in appearance, having long slender cylindrical bodies tapering at each end and covered with a tough skin with the mouth at the tip, but they do not have a well-defined head. Among plant parasites is the sugar beet eelworm which caused havoc on the Continent as far back as 1859. It was not recorded in Britain until 1928, but is now found wherever the beet is grown. The attack causes plants to become stunted and wilted. Another very important parasitic eelworm is the potato cyst nematode. It is a major pest and has caused heavy losses in every country where potatoes are grown. It feeds on the potato whilst in the larval stage and comes out after about two months as an adult. The head of the female remains in the potato while the rest of the body protrudes. After fertilization the female swells up, as she contains several hundred eggs. Then she dies, and her remains form a hard case round the collection of eggs. The case is known as a cyst. The eggs will more readily hatch in the presence of potato and related plants and can lie dormant for ten years or more.

Man and his insect pests

When man first evolved from his monkey-like ancestors some half a million years ago he took his personal insect pests, fleas, lice, bugs and so on with him. When he lost his hair and started wearing furs and moving his family into caves for warmth and protection his pests too appreciated the improved conditions and they, like man, thrived accordingly. The pests have remained loyal to man ever since and it is only in recent times that, with the aid of pesticides, he has begun to reject them. Although throughout the ages these pests have caused man acute discomfort, this is only a minor part of the story because, more importantly, many of them have been and still are carriers of virulent organisms. During the Thirty Years War (1618–48), plague (carried by the rat flea), dysentery (transmitted by, among others, the housefly) and typhus (carried by the human louse) were said to be together responsible for the death of ten million people compared with the 350,000 who were killed in actual warfare. The bacterium that causes plague affects rats and man equally severely and is conveyed from one to the other in the saliva of the small wingless flea. The fleas leave the bodies of the rats when they die of plague and migrate to man. In medieval times practically no house was free of rats and the great epidemics of plague killed millions of people. Between the years 1345 and 1351 plague (or, as it was better known, the Black Death) killed more than 25 million people in Europe out of a total population of 100 million.

Typhus fever, caused by a virus-like organism called *Rickettsia* carried by the louse, has slain soldiers and civilians by the million. The disease is not transmitted by louse bites but by the entry of infected louse excrement and body fluids into abrasions in the skin. The squashing of lice and the scratching associated with it often led to infections. Until quite recent times jails crowded with people awaiting trial were subject to dreadful epidemics of

7

typhus—indeed it was known as 'jail-fever'. Judges were as liable to be infected as others and the shaking of their wigs to free them of lice constituted a grave danger for other unfortunates standing or sitting below. The risk of typhus was recognized—though no one associated it with the lice—and judges were provided with a little bunch of flowers ('nosegays') whose scent was supposed to keep the 'evil humours' of the disease away. But there was no stigma attached to being lousy. Quite the opposite. Absence of lice from one's body was regarded as a sign of lack of virility. In medieval times the sanctity of holy men was enhanced in proportion to their lousiness. It is recorded that when the body of Thomas à Becket was disrobed after his murder the lice in his hair-cloth garment, in the words of a contemporary chronicler 'boiled over like water in a simmering cauldron'. Far from being disgusted or afraid the onlookers were overcome with 'the joy of having found such a saint'. To turn to more modern times, during the First World War typhus killed more people than did the guns of the combatants. A mild form known as 'trench-fever' (which was seldom fatal) was common among the louse-ridden soldiers of the European trenches. It disappeared after the war.

Other insects, some mosquitoes for example, while not actually living in such close personal proximity to man as fleas and lice, have still found him good to feed on. Soon after leaving their pupal skins the adults mate, after which the malaria-carrying mosquito must have a feed of human blood before her eggs will develop. In feeding she may pass the malarial parasite into the human blood stream. Malaria has been described as the greatest killer of all time. Among those who have died of the disease have been many famous people—Alexander the Great aged 32 in 336 B.C., Oliver Cromwell aged 59 in 1658, and Lord Byron aged 36 in 1824. In Rome, popes and cardinals died in staggering numbers: at one time, the mortality was so high that (for this and other reasons) the papal seat was moved from Rome to Avignon.

In the seventeenth century the disease was rampant in Europe (and England) and both King Charles II and the Dauphin of France were cured by the 'Jesuits' bark' (the bark of the quinine tree which was first brought to Europe by the Jesuits). Oliver Cromwell refused the Jesuits' bark and died—a strong Protestant—of malaria. The fall of the Greek and the Roman civilizations has been partially attributed by some historians to malaria.

In Rome the population made 'fever' a goddess in an attempt to placate her, and sanctuaries were built to her. To little avail.

Although it is of interest to read of the effects of malaria in times past we must never forget that it is still very much with us, and indeed it exists as a threat to more than one thousand million people throughout the world. It reached one of its peaks in the 1940s when it was calculated that there were more than 350 million cases of malaria throughout the world, of which about 1 per cent or about 3 million people died annually. In recent years there has been a marked reduction down to 100 million cases (largely due, as we shall see in Chapter 4, to the use of DDT).

At one period during the Second World War it was estimated that the probability of a soldier being incapacitated before he engaged the human enemy was over ninety per cent. Most of the casualties were due to insect-borne diseases.

As indicated earlier in this chapter, man's personal insect pests are not the only ones that he has had to face in his long struggle for survival. The attacks upon his food plants by insects have been almost as fatal, and the speed of spread and the scale of some of these attacks is staggering. The Bible speaks of the great plagues of locusts that stripped areas bare of all vegetation and today the same locust is a potential pest over eleven million square miles of southern Europe, Africa and Asia, ranging from southern Spain and Asia Minor, the whole of northern Africa, through Iran, Bangladesh and India. Within this area more than 300 million people are liable to suffer devastation of their crops. The locust is a strange insect: for a number of years there may be so few of them that it is almost impossible to find any for experimental work. During this phase it is almost solitary in its habits. Then there are sudden and unpredictable increases in local populations followed by mass eruption. Swarms may comprise up to 1,000 million individuals capable of consuming some 3,000 tons of food in a day, destroying cultivated crops and plantations in their way, devouring everything green. Plagues are eventually brought to an end by bad weather. The few survivors revert to the solitary phase in which they are relatively harmless: between plagues lasting six years or more there are equally long recessions when only solitary locusts are found. The last plague ran from 1950 to 1962. There was a threat of one in 1967 which by 1968 looked serious, but by mid 1969 it had receded.

2. WEEDS

To the 'pure' biologist there is no such thing as a weed. All plants are plants in their own right to be observed and studied with more or less interest according to his inclinations and interests. The 'applied' biologist is not so democratic. In his eyes all plants are not equal. To him there are 'good' plants and 'bad' plants. The good yield food, or other useful products, and are to be encouraged; the bad interfere with this food production, and are to be discouraged. Into the latter category come weeds. They may be defined as 'plants growing in the wrong places'; in other words the question of whether a plant is a weed or not is a subjective judgement. In addition however, successful weeds have in common certain characteristics responsible for their success. They tend to be aggressive, competitive and adaptable. Their most important attributes are efficient reproduction combined with mechanisms that permit survival under temporarily unfavourable conditions.

Weeds are harmful to crops in many ways:

(1) They compete for water.

(2) They compete for light, especially those with large coarse leaves which tend to shade the crop plants.

(3) They compete for nutrients; thus a plant of yellow mustard needs twice as much nitrogen, twice as much phosphoric acid, four times as much potassium and four times as much water as a well-developed oat plant.

(4) They compete for space above and below ground. Crops such as lucerne are slow to form a ground cover and may be overgrown by weeds before they have the chance to become properly established. Some weeds such as bindweed and cleavers even use the young crop plants for support. Competition takes place below ground as well as above it and numerous experiments have shown that the roots of crop plants such as wheat may be reduced by two-thirds when grown in the presence of wild oats.

(5) Some weeds such as dodder are almost completely parasitic, absorbing water, mineral salts and carbohydrates from the crop plants which they entwine. Others such as yellow rattle, red rattle and eyebright attach their roots to the roots of grasses and draw nutrients from them.

(6) Weeds may reduce the value of produce. Weed seeds may

increase the cost of cleaning and some are extremely difficult to remove from crop seeds. Where seeds are being used as foodstuffs there is the possibility of 'tainting'—e.g. wild onion bulbils give an onion taint to the flour and render it worthless. Some weed seeds such as darnel and corncockle are poisonous, but they are eliminated by seed cleaning.

(7) Weeds may be poisonous to stock. Grazing animals are normally selective in their feeding and in most cases avoid poisonous plants; but nevertheless many cases of poisoning do occur, particularly by water dropwort and ragwort, especially after the latter has been sprayed with a herbicide which seems to make it more palatable. It is perhaps of interest to note here that sometimes a cow may become 'addicted' to a poisonous plant. After having been poisoned, it may be cured by the vet but on being returned to the field it may make a bee-line for the same poisonous plant.

(8) Weeds increase the cost of growing crops because they increase the difficulty of harvesting.

(9) Weeds may harbour pests and diseases, e.g. dock plants can harbour a sawfly which will attack fruits in orchards; and the organism which causes club-root disease in cabbages, cauliflowers and turnips may also attack certain weeds related to these crop plants and thus form a focus for new infections.

(10) Even in areas not under crop plants, weeds may be harmful: they can for example block drainage ditches and canals and decrease the amenity value of stretches of water.

Reproduction of weeds
There are four main groups of weeds, classified according to their life histories. They are:

(1) EPHEMERALS such as groundsel, which complete their life cycle from seed to seed production in a remarkably short time and thus produce many generations within any one growing season.

(2) ANNUALS which start from seed in spring or summer, develop into mature plants, ripen their seed and die in the same season. There is a long list of important weeds in this group such as fat hen, cornflower, fumitory, corn poppy, bindweed, white mustard, black nightshade.

(3) BIENNIALS start from seed in the spring and produce a rosette of leaves from a tap root in the first year. In the following

spring an overwintered tap root sends forth a flowering shoot which ripens seed and then dies at the end of the season. Examples are marsh thistle, spear thistle, wild carrot.

(4) PERENNIALS—the aerial parts of perennial weeds usually die at the end of each season, but the underground parts continue to live and they send up new shoots each year. There are a great many important weeds in this group too; and because of their underground organs they are particularly difficult to eradicate. Among them are bishopweed, couchgrass, rosebay willowherb, rushes, plantains, bracken, docks, dandelion, coltsfoot and nettles.

One of the most important attributes of weeds is their ability to reproduce freely. Reproduction may be by seed or by vegetative means.

Reproduction by seeds. Seeds are formed as a result of the fusion of male and female elements in the flower, and they generally carry a reserve of food materials. Most common weed species reproduce by seeds at some stage in their life cycle (bracken and horsetails are exceptional in that they produce spores which, unlike seeds, are not produced by sexual fusion and have little, if any, reserve food materials). Some weeds such as chickweed and gorse can be seen in bloom at any time of the year because their flowering is independent of day-length. The spring-flowering species, i.e. dandelion, are those which require medium and increasing day length for flower initiation; whereas summer-flowering species such as wild oats need long days, and autumn-flowering species such as the autumnal hawkbit require medium and decreasing day-length. Most common weeds are self-fertile and in some, i.e. dandelion, seeds can be produced without fertilization of the ovules. Some weeds can produce enormous numbers of seeds per plant, e.g. rosebay willowherb—76,000; ragwort —63,000; rushes—50,000; field poppy—17,000. The significance of these figures is best appreciated when one considers that the multiplication rate for a cereal is commonly 25–30 times, grasses 30–80 times, white clover 40–70 times.

From the practical point of view, however, seed production by the whole weed population is more important than that of single plants. One dense weed stand has been estimated to produce more than 5,000 million seeds per acre. Seed production by a dense stand of rushes has been estimated at 8 million per square yard. In the presence of a crop, production of weed seeds is usually much less but is still considerable. In a crop of winter wheat 25 million weed seeds per acre were found at harvest. To appreciate the full significance of these figures it is helpful to remember that the approximate number of seeds in $1\frac{1}{4}$ cwt of barley (the amount commonly sown on an acre) is 1·6 million.

Seeds are of course of no significance unless they can germinate, i.e. unless they are viable. Viability is affected by a number of factors one of the most important of which is *dormancy*. Many weed seeds do not germinate immediately but may have a short or long period before they do so. The seeds of blackgrass are dormant for only a few weeks whereas those of the common vetch may remain in this condition for a few years. There are two types of dormancy: (a) *innate* and (b) *enforced*. Innate dormancy is due to features of the seed itself. For example it may be due to the fact that water cannot easily penetrate the seed coat (in which case dormancy may be overcome by cracking the coat), as in various clovers; or the embryo may be underdeveloped, as in hogweed, so that the species requires a period of after-ripening often at low temperature.

Enforced dormancy, on the other hand, is due to external circumstances and as soon as the seeds are freed from these conditions they will germinate. Thus seeds buried in undisturbed soil or under grass for many years will germinate when the soil is disturbed. They have been kept dormant due to the high concentration of carbon dioxide in the deep layers of the soil. Disturbing the soil releases the carbon dioxide, increases the concentration of oxygen getting at the seeds and allows them to germinate. Some seeds may lie dormant in the soil for fifty years or more.

Methods of spread of seeds. Seeds are spread by a variety of agencies, the most important probably being wind—we find that some weed seeds are specially modified for this purpose. For example the seeds of dandelion, groundsel, coltsfoot and others have special 'parachute' mechanisms while others such as docks may have 'wings'. An interesting example of wind dispersal is provided by the tumbleweeds of America. These are much-branched plants which, when mature, are broken off at the base by the wind and are blown along, scattering their seeds as they go. Many types of weed seeds are also carried in the waters of ditches, irrigation channels and small rivers and some of these such as the sedges have bladders which keep them afloat. Observations made on a 12-foot ditch have indicated that in a flood period several million weed seeds may pass any given spot in the course of 24 hours.

Animals of all kinds serve to spread many weeds. The hooked fruits of cleavers and wild carrot are admirably suited for attachment to animal fur or tweed suits of man. Sir Edward Salisbury, the famous botanist, examined the turnups of his trousers after a country walk and found that he had a yield which on planting gave rise to more than 300 seedlings.

The seeds or fruits of many weeds can pass through the alimentary tract of animals completely unharmed. As far back as 1911 an experimenter called Harrison estimated that a cow ate in her fodder, in a single day, around 89,000 seeds of plantain and over 500,000 seeds of chamomile and found that 85,000 and 198,000 respectively were voided in the cow's droppings with germination abilities of 58 and 27 per cent.

Vegetative reproduction
Many perennials, in addition to reproducing by seed, have vegetative methods of reproduction. Some such as the bulbous buttercup have swollen, upright underground stems termed corms. Many have creeping stems above ground (e.g. creeping buttercup) or below ground (like bracken and couchgrass). Some have branched tap roots (i.e. hemlock and docks) whilst others produce bulbs below ground (wild onion) or little bulbs—bulbils —in place of seeds above ground (lesser celandine). Several common water weeds such as Canadian water weed and frogbit, which rarely if ever produce seed in Britain, overwinter by means of turions—vegetative buds which become detached in autumn, lie in the bottom mud and grow out in late spring to form new plants when the water becomes warmer.

Man and his weed pests
Weeds have been and are major pests and some, given the right conditions, can thrive at an astonishing rate. For example the prickly pear was introduced into Australia in the late eighteenth century as a decorative potted plant. It 'escaped' and by 1925 when a survey was carried out it had covered 60 million acres and was spreading through the country at a rate of about one million acres per year. In water, the floating fern invaded the Kariba Dam in 1959 and in less than one year had covered an area of 75 square miles; in 1½ years an area of 175 square miles was covered,

making fishing and the movement of craft impossible. Another weed, the water hyacinth, entered the southern swamps of the Nile early in 1957. When first reported in 1958 it had already spread along 1,000 kilometres of the river at a rate of more than two kilometres a day—interfering with the water supply, blocking channels, delaying transport and providing breeding places for mosquitoes.

(3) THE FUNGI

The fungi are, in a number of ways, rather a curious group of plants. Although many of them such as mushrooms and toadstools can be readily seen with the naked eye they are really built up of microscopically thin threads termed hyphae. Although they can attain some complexity of structures, as in the examples mentioned above, one of their main features is the rather negative one that the plant body is not divided up into root, stem and leaf. This negative feature they share with their relatives the algae but, unlike the algae and almost all higher plants, the fungi do not possess the green colouring matter chlorophyll (otherwise so characteristic of the plant kingdom) and as a consequence they cannot utilize sunlight to manufacture sugars and other foodstuffs. They must therefore live as do animals and man on foodstuffs that have been already made for them either by green plants or by animals feeding directly or indirectly on these green plants or their products.

Fungi may live on dead or non-living organic matter, in which case they are termed saprophytes, or on other living organisms in which case they are known as parasites. The saprophytes play a very important role in nature's economy, particularly in the soil where—together with bacteria—they break down dead plants and animals into simpler substances which become available for use again by plants. Some saprophytes are utilized by man, thus yeasts and other fungi are used in brewing processes; some such as *Penicillium* produce antibiotics, and others are used in various industrial and food-manufacturing processes. The parasitic fungi on the other hand are harmful in that they induce disease—a deviation from the normal physiological condition. In some cases the parasite may kill its host quickly and then continue to live on the remains as a saprophyte. This is considered to be a fairly primitive form of parasitism since the parasite opens up its 'prey' to all kinds of saprophytes with which it is then in competition. In the more highly developed forms of parasitism the parasite keeps its host alive and continues to draw sustenance from it. An example of this type is the fungus causing smut disease of various cereals. The fungus grows through the tissues of the cereal doing little or no harm until it produces its black spores (or 'smuts') in the flowers, thus replacing the seeds.

Possibly the highest form of parasitism is symbiosis where the fungus and the plant on or in which it lives appear to have developed an almost perfect balance in which each partner derives benefit from the other. Lichen, which is made up of a fungus and an alga growing together, provides an excellent example of symbiosis.

It has been estimated that there are more than 200,000 species of fungi and some 1,000–2,000 new species are described annually. Reproduction is by means of spores—small particles of fungal living matter, covered by a protective coat, which may be produced by a sexual process (the details of which vary considerably from species to species) or asexually. Classification of the fungi is to some extent based on how the sexually produced spores are carried. For example members of one of the major groups, the *Basidiomycetes*, bear the spores on little clubs or basidia. This group includes most of our mushrooms, toadstools, bracket fungi (many of which destroy millions of pounds' worth of timber and lumber annually) and puffballs, but also some microscopic fungi such as the rusts and smuts which are important parasites of, among other plants, the cereal crops. The Romans, by the way, recognizing the importance of the rust disease, organized the *Robigalia*—a festival designed to propitiate the rust god Robigus in the hope that he would avert the effect of rust on their crops. Rust diseases caused by various species of *Puccinia* are still important today. In the United States alone losses due to this disease of hundreds of millions of bushels of wheat are recorded each year. Some of the rust fungi spend part of their life cycle on one type of plant and part on another: black rust, for example, attacks cereal crops but overwinters on the barberry bush, where it produces a type of spore capable of attacking cereal plants.

The *Ascomycetes* carry their sexually produced spores (generally eight in number) inside small sacs or asci (singular ascus).

This group contains many fungi which are very important in agriculture, medicine and industry. Many important plant diseases are caused by ascomycetes, among them apple scab, brown rot of stone fruits, powdery mildew. *Endothia parasitica*, a fungus from the East, has completely destroyed the American chestnut and is now threatening the European chestnut. Another fungus, *Ceratostomella ulmi*, which is carried by a bark beetle, is relentlessly destroying the elm tree in Britain.

One very interesting ascomycete is *Claviceps purpurea* which attacks the flower heads of grasses and cereals—particularly rye—and replaces the seed with hard, blackened fungal tissue termed sclerotia or 'ergots'. If these sclerotia are not detected and are milled with the grain, the results can be disastrous. Throughout the Middle Ages we hear of 'St. Anthony's Fire'—victims suffered from hallucinations, including one in which they thought themselves to be aflame. These hallucinations were caused by people eating ergot-infested flour. Another manifestation of ergotism was gangrene: the toxins in the ergot interfered with the blood supply, and the complete loss of limbs was not uncommon. The last outbreak of the disease in Britain was recorded in the 1920s in Manchester, but only a few years ago there was a distressing outbreak in France. Ergots are now used medicinally to prevent excessive bleeding; and artificial inoculation of rye heads is carried out in some European countries.

The third major group of the fungi are the *Phycomycetes*, which includes rather a miscellaneous lot. The only thing which they appear to have in common is that the hyphae consist of long hollow tubes, i.e. they are not divided up by cross walls as are the Basidiomycetes and Ascomycetes groups. The Phycomycetes contain a number of important plant pathogens including some such as *Pythium* (which attacks and destroys young seedlings), *Phytophthora infestans* (the cause of potato blight), and *Synchytrium endobioticum*, which by causing the disease known as wart threatened destruction of the potato crop but which is now little more than a laboratory curiosity since plant breeders have developed resistant varieties of potato.

The fourth and last major group are known as the *Fungi Imperfecti* because they do not have any method of sexual reproduction. They reproduce almost exclusively by means of asexual spores which are produced in great variety and in enormous numbers. The group includes a great many important plant parasites.

17

Distribution of fungi is mainly by means of spores, which are produced in enormous numbers. A giant puffball has been estimated to contain 7,000,000,000,000 spores. The large bracket fungus *Ganoderma applanata* may liberate 30,000,000,000 spores a day, apparently maintaining this output daily for six months. In the stinking smut of wheat a single diseased grain may contain over 12,000,000 spores. The main agent of spread is the wind and some spores may be transported very long distances; for example, there is good evidence from Canada that under certain conditions the spores of black rust of wheat can be carried 500 miles or more. This is obviously a major factor to be taken into account in determining control measures. Spores of some fungi such as *Aspergillus* and *Penicillium* have been found at a height of 20,000 feet although they tend to be scarce beyond 11,000 feet. Spores of rust fungus trapped at 7,000 feet have been found to germinate readily.

Fungal spores may retain their viability for long periods, e.g. dry spore dust of *Aspergillus oryzae* up to 35 years and *Ustilago crameri* up to 64 years; and they show amazing powers of resistance to extremes of heat and cold. Sealed in vacuum tubes, slowly-dried spores of some fungi have withstood the temperature of liquid air (− 190°C) for 77 hours. Some spores can withstand the direct action of boiling water or live steam for considerable periods and some moulds will grow at a temperature of − 6°C. Some spores are spread mainly with the seeds of the plants on which they grow. This is the chief method of spread of bunt of wheat. Insects are also connected with the spread of some spores.

As well as reproducing by spores, some fungi reproduce by vegetative means. The walls of the hyphae of the fungus causing leaf-stripe of oats become thickened and darkened, and can tolerate adverse conditions. Much larger and more complex masses of vegetative hyphae may become united into 'sclerotia' which may measure several inches across and are as hard and as heavy as a stone. Small sclerotia are also known and among the commonest are those of the fungus causing black scurf of potatoes. Some fungi, such as the dry-rot fungus and the honey fungus which attacks trees, produce thick strands of closely intertwined hyphae known as rhizomorphs. These have been likened to shoe-laces but they are often much branched. They are remarkably tough and resistant—those of the dry-rot fungus have been found

18

passing through the mortar of walls to reach fresh structural woodwork on the other side.

Nutrition
Parasitic fungi reach their food by sending their hyphae into the interior of the host plant, or by remaining for the most part superficial and sending only specialized hyphae known as haustoria into the surface tissue (as in many of the powdery mildews). The parasite may enter through natural openings such as stomata (pores in the leaves) and lenticels (pores in the bark), through the flower, through wounds or through root hairs. The fungi possess a vast array of enzymes which may be excreted by them into the host to dissolve a way ahead or to make foodstuffs more readily assimilable.

Man and his fungal pests
Some fungi attack animals and man, causing for example ringworm or inflammatory conditions of the lung, but their main attack is against plants. Some plant diseases have reached epidemic proportions—e.g. the potato blight which swept through Europe in the 1840s; the coffee-leaf disease which devastated the flourishing plantation industry of Ceylon and almost halted its development elsewhere in the East between 1869 and 1883; the vine mildew which spread from France through southern Europe from 1878 onwards bringing panic and ruin to growers everywhere in its track; and the rind disease which almost brought about the destruction of the sugar-cane industry of the West Indies in its peak years about 1895.

One of the best known plant diseases is the potato blight (caused by the fungus *Phytophthora infestans*) which struck with such devastating force in Ireland in the 1840s. Almost overnight the green potato fields of Ireland were reduced to rotting, putrefying masses. The first warnings appeared in the editorial columns of the *Gardner's Chronicle and Agricultural Gazette*, August 23rd, 1845:

A fatal malady has broken out amongst the potato crop. On all sides we hear of the destruction. In Belgium the fields are said to have been completely desolated. There is hardly a sound sample in Covent Garden Market. The disease consists of a

19

gradual decay of the leaves and stem which become a putrid mass and the tubers are affected in a similar way.

The disease was widespread in Europe, being reported in England, France, Germany, Belgium, Poland and in particular Ireland. 'We stop the press with very great regret, to announce that the potato disease has declared itself in Ireland. The crops about Dublin are suddenly perishing. Where will Ireland be in the event of a universal potato rot?' (*Gardner's Chronicle and Agricultural Gazette*, September 13th, 1845).

The failure of the potato crop caused dire distress and famine throughout Europe because it was the basic diet of the poor; but alternatives such as wheat, oats, rye, though comparatively expensive, were available. And cereals were not affected by the blight. But in Ireland the potato was practically the only food crop grown. Its pitiless destruction by the fungus left the poor facing famine on a scale unknown since Biblical times. Thousands of people perished in that terrible winter of 1845–6.

In the spring the few potatoes that had not been eaten or destroyed were planted again in the hope of a harvest. And again the blight struck, in the summer of 1846. The scene is described by a priest of the time:

On July 27th I passed from Cork to Dublin and this plant blossomed in all the luxuriance of an abundant harvest. Returning on August 3rd (one week later) I beheld with sorrow one wide waste of putrefying vegetation. In many places the wretched people were seated on the fences of their decaying gardens wringing their hands and wailing bitterly at the destruction that had left them foodless.

The human spirit could stand no more. In the fifteen years from 1845 to 1860 more than a million men, women and children died of hunger, disease and pestilence. 1½ million emigrated. Those who could afford it went to Canada and the United States. The poorer went to the ports of Glasgow and Liverpool, and travelled under conditions of indescribable horror—standing all the way because there was no room to sit down, many dying *en route* without their companions realizing because the dead were still standing erect, held up by the crush of their fellow passengers. The potato blight struck a fearful blow at Ireland and its scars are not yet healed.

Note: Insects, weeds and fungi are not the only pests with which man has to compete. Among others are some of his fellow back-boned creatures. These are dealt with in Chapter 8.

Chapter 2

METHODS OF
PEST CONTROL

In the course of his struggle, man has used a great variety of methods to try to control pests. These include mechanical devices, special cultivation, flood, drainage, oiling, quarantine, eradication, the use of resistant varieties, biological methods and chemical methods.

Mechanical methods
These are among the oldest methods used by man. He would pick his 'personal' parasites such as fleas and lice from his person, and a fly swatter would probably be one of his earlier inventions. When he started cultivating crop plants he pulled out and discarded weeds or plants which were obviously diseased. Later he developed tools which enabled him to tackle the more deeply-rooted weeds, and thus digging implements evolved which in time led to the plough.

One of the objects of cultivation is to turn over the top soil so that plant remains are buried and a clean seedbed is provided for the new crop. This is one of the major means that the farmer has had through the centuries of preventing weeds from competing with his crop. The plough is the traditional tool for effecting such cultivations, dating back to pre-Roman times.

The first ploughs were made of wood, but an improvement was effected by plating the wood with iron in the eighteenth century. Wood was subsequently replaced completely by cast iron and in more recent times by high quality steel. The heavy ploughs of the Middle Ages needed eight oxen to draw them through the unyielding clay soil, but changes in design during the eighteenth and nineteenth centuries meant that ploughs were developed which could be pulled by four-horse or even two-horse teams. The Industrial Revolution led to the emergence of the steam plough, but the major change came with the development of the tractor in the twentieth century. There are some who feel that with the development of herbicides the days of the plough may be numbered.

Harrows and cultivators are relatively recent developments. They are used primarily to form a good seedbed with fine soil, but they also clear the soil of roots and underground stems by dragging them to the surface. By disturbing the upper layers of the soil they also bring weed seeds to the surface and cause them to germinate. This is advantageous to the farmer in that the developing young weed plants can then be easily destroyed. The rotary cultivator is particularly valuable in that it cuts up the roots and underground stems of perennial weeds and causes them to dry out.

Ploughs, harrows and cultivators are all very important implements for disturbing the soil surface; but they can in general be used only when the ground has no crop growing on it, say in winter, early spring or late autumn. The hoe, on the other hand, is much more selective and because if its accuracy it can be used in the growing crop. On a small scale, and with plenty of labour available, the hand hoe was a very efficient tool—particularly after the introduction of steel in the mid nineteenth century. This was followed by the horse-hoe in which an assembly of blades was drawn along. This was replaced by tractor-mounted steerage hoes in the 1950s, some of which had independent hydraulic control for undulating ground. However, as with the plough—and perhaps even more so—the value of the hoe has decreased in importance with the development of herbicides.

Machines have been developed to deal specially with trees, tree stumps and the like. Winches of various kinds are available for attaching to tractors and these can pull down large trees. For removal of roots that might re-sprout, a machine is used which cuts the roots below the surface; the cutters are V-shaped blades which are mounted on a tractor. Cutters are also available which will cut off trees up to 20 inches in diameter flush with the ground whilst travelling at speeds of two to three miles per hour. Anchor chains can also be pulled along strung between two tractors. The chain knocks down the trees as it comes into contact with them, provided the tractors have sufficient power.

The keeping of roadside verges down to reasonable heights is a constant problem. Hand-scything is effective but slow and nowadays control is achieved mostly by machine mowing or by the use of chemicals.

Rotation

The introduction of turnips, red clover and potatoes into Britain in the seventeenth century allowed the farmer not only to vary his and his customers' diet but it also meant that he could ring the changes on the crops grown in any one field. Before this time, the growing of cereals interspersed with fallows (a season when the land was allowed to 'rest' and no crop was taken) was the common practice in Britain. The yields were low and the cereals were blasted by disease and insect attack. The introduction of new crops made rotation possible, so that one year the farmer might grow potatoes, the next year barley, the next clover or beans, and the fourth year wheat and then repeat the cycle. There were a number of variations on the rotation theme but all were beneficial to the soil and to the crop.

Each crop has its own particular mineral requirements, so the soil was not robbed by a succession of similar crops. The development of chemical fertilizers in the mid nineteenth century was an added bonus. Also, each crop has its own particular weeds and diseases, so that by rotating the crops there was no continuous build-up of specific weeds and diseases year after year. Potatoes were known as a good 'clearing' crop—because of their fast-growing canopy of leaves and the cultivations in the growing crop. Even in the late 1940s, the rotation methods of the nineteenth and twentieth centuries were still in use; but with the massive reduction in the labour force and the development of chemical pesticides, some farmers no longer follow set rotations and changes are made only when disease or uncontrollable grass weeds force a change.

Removal of weed seeds

The cleaning of threshed seed by movement of the air dates from early times. The most primitive form, which is still used in some underdeveloped countries, is winnowing. This consists of throwing the seed in the air so that wind currents carry away the chaff and light weed seeds leaving the heavier crop seeds to fall to the ground. In Britain, manually operated fans and hand winnowers had been developed by the late eighteenth century and were widely used. Later still, winnowers were incorporated into the threshing machines.

In the 1920s a number of new techniques were developed. One was used in the main for getting rid of dodder seeds from the mixture. Dodder can be a very troublesome weed as the plant becomes parasitic on the crop plant and draws sustenance from it. Although dodder seeds are about the same size as the clover seeds from which it was desired to separate them, they differ in that their coats are very rough whereas the coats of the clover seeds are smooth. Iron filings were introduced into the seed mixture; these attached themselves in the crevices of the rough-coated dodder seeds but slid off the clover seeds. A powerful magnet was then passed over the seed mixture. This attracted the iron filings and the attached dodder seeds, leaving the clover seeds behind. This technique played a very important part in controlling the distribution of dodder seeds.

Gravity separators (originally developed for separating out different grades of ore) have also been employed. They use differences in the weights of different seeds to effect separation. Another interesting device used for separating minerals has also been adapted for use with crops; this is the electrostatic separator which exploits differences in the conducting properties of seeds. Photo-electric cells, too, have been employed which separate seeds on the basis of their colour. The importance of keeping weed seeds out of mixtures can be gauged from the variety and sophistication of the devices described above.

There is now legislation to control the amounts and kinds of weed seeds in crop seed. The Seeds Act of 1920 was the beginning; new seed regulations were introduced in 1961 and powers for the control of weed seed dissemination in crop seed are contained in the Plant Varieties and Seed Act 1964. Under the Act, the seller of seeds (or seed potatoes) must give the buyer particulars of variety, purity and germination either on or before sale. Wild oats, dodder, docks and sorrels, blackgrass and couchgrass are defined as injurious weeds and it is required that the names and number of seeds of these weeds present in a given size of sample of certain crop seeds should be specified. In addition the percentage of weed seeds present in these crop seeds must be declared if in excess of 0·5 per cent by weight.

Under the Weeds Act of 1959 the Minister of Agriculture (or his equivalent) has the power to require an occupier of land to cut down or destroy 'injurious weeds' (spear thistle, curled-leaf dock, broad-leaved dock and ragwort) within a specified time. The Minister may also enter land and take the necessary action, recovering the cost of doing so from the occupier. The idea behind the legislation is to prevent the spread of these weeds on to the land of neighbouring farmers.

Flooding

Flooding is a very old technique used for the control of weeds in the paddy fields of the East, but it has also been used particularly in the United States for the control of perennial weeds in other situations. In California, for example, some particularly difficult areas may be surrounded by dykes and the whole area then submerged under 10 inches of water for several weeks. In this way a number of difficult weeds have been eradicated. The practice is also said to reduce the numbers of eelworms and pathogenic fungi.

Drainage

In the control of the malaria-carrying mosquito much can be done by drainage. Both its larval and its pupal stages are spent in water and the draining of unhealthy towns and the filling in of stagnant, swampy pools are very effective measures against it. In England, Holland, France, Italy, Algeria, America and many other places, enormous tracts of country which were formerly useless and pestilential have been rendered healthy and productive by such means. The draining of the Pontine Marshes in the 1930s helped to free Rome from the grip of malaria.

Smaller areas of water, if overlooked, may also constitute a grave danger by providing breeding grounds for the mosquito. A sagging gutter may hold enough water to support large numbers of larvae and even holes in rotten trees may harbour the larvae and pupae.

Oiling

Oiling kills mosquito larvae by suffocation and by its toxic action. It enters the spiracles and breathing tubes of the larvae and suffocates them. Crude liquid petroleum is said to be one of the best agents, because it spreads so effectively on the surface of water. Oils are generally applied as sprays, but other ingenious methods have been devised. For example, in some countries buffaloes may be made use of by oiling them at night so that when they wallow the next day they spread the oil.

Fire
Fire is another very old method of reducing weed growth and it is still used in certain situations. For example, flame throwers using propane or butane gas have been used to remove unwanted woody vegetation such as scrub land before cultivation. The mineral salts released from the burning plants are excellent fertilizers in the soil.

In some countries it is common practice to 'fire' the land, thus destroying old vegetation and allowing new growth to come through. Not only does the firing get rid of old vegetation but it may also stimulate seeds in the ground to germinate. The technique is used on grassland; and in Scotland the burning of heather moors is commonly practised to get rid of the old woody shoots which are inedible by grouse and sheep, and to encourage the new growth of young, fresh, green heather plants.

Fire in the form of blow lamps is also used to eliminate bugs and other insects from crevices in the stone walls of houses, outhouses and stables.

Quarantine
Many of the worst outbreaks of plant, animal and human diseases have come about through the introduction of parasites from some other part of the world into a country where they were hitherto unknown. The movement of people, animals and plants is probably one of the major factors in this spread and most countries have now established quarantine regulations allowing the import of plants and animals, and sometimes of people, only under certain strictly specified conditions. At times there may be a complete embargo; at others there may be a period of detention required, the material being held for a certain time in order to ensure that it is disease-free. This is the system that is employed in the detention of dogs from abroad in order to prevent the spread of rabies. Humans who have been in close contact with certain diseases may be kept under very close observation and certificates

of vaccination may be needed. In the case of plant material this is sometimes admitted subject to disinfection on entry. Disinfection is most frequently prescribed for seeds but it is also used with planting stock, particularly when it is in a dormant condition.

Quarantine has no doubt been very effective in preventing the entry of many alien insects and diseases; yet, in spite of the regulations, scarcely a year passes without some new pest being reported. For example, despite the strictest precautions and a good deal of publicity, single Colorado beetles are still reported in Britain from year to year. The United States probably has the most stringent and strictly applied quarantine measures in the world and yet very important plant diseases have evaded them. Dutch Elm disease, caused by a fungus (*Ceratostomella ulmi*) and carried by a little insect, the elm bark beetle, is one of the deadliest diseases of the elm tree, and was imported into the States in logs from Holland from 1928 onwards. It reached several of the eastern states before a concerted effort was made, costing $5m. from Federal funds alone, to check its spread. In 1938 the ring spot disease of potatoes was first seen in Maine whence it spread with extraordinary rapidity through the potato-growing areas of the country.

Most other countries have had similar experiences. They may be attributable to a number of causes—the difficulty in recognizing disease in a plant; the increased traffic in plants and plant parts by traders and by scientists; the speed-up in transportation so that latent diseases have no time to develop; and the lack of knowledge of how certain diseases may be spread.

The disruption of trade caused by quarantine has led to efforts being made to secure relaxation by measures designed to guarantee that only certified disease-free stock will be offered by the exporting country, so that certificates may be issued by properly accredited authorities in such exporting countries. In Britain it is against the law to import plants intended for planting unless they are accompanied by a certificate from the plant disease service of the country in which they were grown. The certificate must vouch that the living plants were thoroughly examined (on a date which should not be more than 14 days prior to shipment) and were found to be healthy, and must also state that no plant of elm or sugar beet is included. This latter requirement is with a view to checking any further introduction of Dutch Elm disease and to check virus diseases of the sugar beet crop. For all potatoes an additional certificate is required that no case of wart disease (see page 32) has occurred on the farm on which they were grown, or within two kilometres of it. Any consignment can be inspected and treated, destroyed or returned, if it does not meet the stringent requirements.

Notification
Many countries list certain plant diseases which must be notified by the grower to the authorities. This is usually prescribed for diseases which have a limited distribution or against which restrictive measures are in force (e.g. wart of potato in Britain). By this means it is often possible to limit the spread by swift treatment or by destruction of affected plants.

Eradication
In some countries diseases may be eradicated by the extermination of diseased individuals. Such is the policy employed in Britain against foot and mouth disease of cattle. Not only are infected animals destroyed but also those which have been in contact with diseased animals. Such a slaughter policy is costly—the foot and mouth outbreak in 1967-8 cost the country over £35m.—and cannot always be measured in terms of money. Specially bred herds that have taken many years to build up may be destroyed, by order, in a matter of hours. There have been critics of the system but there is no doubt that it has been very successful in containing the disease.

Another very successful campaign was that waged against the disease of orange, lemon and grapefruit trees—citrus canker—which was introduced into the United States from Japan in 1911 and became well-established in Florida and other fruit-growing areas. So menacing did it become that a policy of eradication was begun and some 4 million grove and nursery trees were deliberately destroyed. By 1927 Florida was cleared of the disease, and by 1935 it was reported that all growing areas were completely cleared. The cost was more than 2½ million dollars. Subsequently, cases were reported in non-commercial areas in Texas and Louisiana and a further 13,600,000 trees were destroyed by the end of 1936. From 1937 to 1940, 1,114 trees were found in the two states but in 1940 only one infected property in each state could be detected, while all the commercial areas in the United States maintained their freedom from the disease.

Another example is concerned with the eradication of the Mediterranean fruit fly in Florida. The pest was first discovered in March 1929 and within two months it was on almost 700 properties; a Federal quarantine was imposed on 1st May and Florida was divided into three zones. In the infected zones, areas within one mile of any spots where the fly was found, all fruit and vegetables susceptible to attack were destroyed and new planting of susceptible plants was forbidden. In the second series of zones, areas within nine miles of the outer boundary of infected zones, the cultivation of fruit and vegetables liable to attack was prohibited between 1st May and 1st November and the movement of such fruits and vegetables was allowed only into such states where, for climatic or other reasons, it was unlikely that the fly could establish itself.

The third, outer, zone included all the rest of Florida. In this zone no mature fruits (except sour limes) were allowed to remain on the trees after 15th June and the export to all other states where the fly might become established was prohibited. At the height of the campaign about four thousand men were employed. The infestation was wiped out by November 1930, the total cost being nearly 5 million dollars. The pest was rediscovered in 1956 and although a Federal programme of eradication was once more put in train it did not need to be as vigorous as in the 1920s, because of the assistance of insecticides.

A very important disease wherever cereals are grown is black rust. It is caused by a fungus, *Puccinia graminis*, which spends its summers attacking cereals and then overwinters on barberry bushes. On the barberry, it spends a part of its life cycle producing spores in spring which can infect the cereal crop. One obvious way of attacking the fungus is to destroy its overwintering home and this policy has been adopted with considerable success in Europe and in the USA. As early as the seventeenth century, before the cause was known, it is said that barberry eradication laws were in force around Rouen because of the obvious part played by the barberry in blasting the corn. In several European countries it is forbidden to have barberries growing within 200 metres of cultivated land, while in others the barberry is completely outlawed. This is the situation in Australia where every occupier of land is required to destroy any barberries growing on it. In the USA great barberry eradication campaigns have been launched over the years. The high point was probably 1936, when

30

68½ million barberry bushes were destroyed in that single year. The eradication of the barberry bush is not the whole answer. The fungus is a versatile one and in some cases it has short-circuited the barberry and moves from cereal to cereal over the seasons. Nevertheless there is no doubt that barberry eradication has saved the grower millions of dollars. One estimate puts the saving in the USA at more than $20m. per annum.

The use of resistant varieties of plants

In 1815 Thomas Knight noticed that in a field of cereals attacked by fungi, some plants were immune and unaffected by the attack. He suggested that it would be worth while to breed from these resistant plants and thus develop 'strains' with this desirable characteristic. This marked the beginning of this method and since then it has met with considerable success. A few examples may be cited.

In the 1880s the sugar-cane industry of Java was seriously threatened by the 'Sereh' disease and might well have succumbed to serious competition from the sugar beet crop in other parts of the world. Fortunately some wild varieties of sugar-cane were discovered which were resistant to the disease and although these were not as high-yielding as the cultivated strains it was possible to cross-breed them with the cultivated strains and breed resistance into the latter. Thus the industry survived.

Another example, this time of a slightly different technique, was used in the grape-growing industry of France. The wine-producers were facing almost total disaster and one of France's oldest and most productive industries looked like going to the wall when the *Phylloxera* insect struck the vines, between 1858 and 1863. From ancient times the only species of vine used for growing wine grapes in France has been *Vitis vinifera*, but this was very susceptible to the little vine pest which had been introduced from America when vine-growers were experimenting with species of vine from that country. At its height, the pest, which attacks the roots and causes them to decay and die, ruined 2½ million acres of vineyards in France. Fortunately the day was saved by importing species of vines from North America which were resistant to the disease. These did not produce desirable grapes but the shoots of *Vitis vinifera* were grafted on to their roots. Since the roots were resistant the vines thrived and produced the original quality grapes from the shoots.

31

Wart disease of potato, caused by a fungus (*Synchytrium endobioticum*), at one time threatened the extermination of the potato until it was noted in 1908 that a few varieties were resistant. These were used for the breeding of new stock and there is now a long list of immune varieties. Indeed, all of the varieties grown commercially today are immune and the disease has become little more than a laboratory curiosity.

There are many successful cases of crop varieties resistant to insect attack. These resistant varieties may occur naturally in the field or they may be produced through the work of the plant breeder introducing resistance into desirable non-resistant varieties.

The hessian fly is a black gnatlike insect, about a quarter of an inch long, which lays its eggs on wheat plants. The larvae hatching from these eggs kill the wheat plant by feeding on the plant juices. Heavy infestations have been known to destroy an entire crop. At one time it caused losses of millions of dollars per annum in the United States alone, but in 1942 a hard red winter-resistant wheat named Pawnee was released for general use. Other varieties are now available. It has been calculated that as a result of their use there has been an annual saving of about $119m.

The wheat-stem sawfly is another major pest which has been hard-hit by resistant varieties. In 1941 it was calculated that losses in Canada alone amounted to several million bushels every year; in North Dakota in the USA the loss was around 3 million bushels valued at about $6m. The adult fly is rather like a black wasp; but its cream-coloured, black-headed larva does the damage, by boring up and down inside the wheat stem and cutting it at the base as the plant matures. Control is difficult because the insect, from egg to adult, spends most of its time within the wheat stem. C.W. Farstand, an entomologist, and A.W. Platt, an agronomist, both of the Canadian Department of Agriculture, noted that solid-stemmed wheat suffered less damage than hollow-stemmed wheat because the sawfly larva could not complete its development in the solid-stemmed plants. By crossing these solid-stemmed varieties (which were not of much agricultural value) with varieties which, though high-yielding, were susceptible to the sawfly, they were able to produce high-yielding, resistant varieties. One variety which was particularly good, appropriately called 'Rescue', was bred between 1944 and 1948, by which time there was enough seed available to plant all the sawfly-infested acreage in Montana, and the damage caused by the fly decreased dramatically by an estimated $4m. in that year.

Thus a conservative estimate for a ten-year period (which is close to the period of usefulness of a variety) would be $40m. for Montana farmers and many times that amount for Canadian farmers.

However, unfortunately for the success of the use of immune varieties, the fungus or insect can sometimes change too and a variety of plant that has been resistant for a long time may suddenly go under. This has not happened in the cases cited above, but a number of fungi are very 'plastic' and can produce new races of fungus almost as fast as the plant breeder can produce new resistant varieties of crop plants. This is particularly true of the rust fungi which attack cereals, the blight fungus which attacks potatoes and the red core fungus which attacks the roots of strawberries.

The nature of resistance is variable, ranging from anatomical features, e.g. hard waxy leaves to complex biochemical factors within the plant. The subject is one of intense investigation because of its considerable practical importance. There are a number of interesting examples of chemical resistance. It has been shown that some oranges and lemons are resistant to the Mediterranean fruit fly because when the female lays her eggs in the fruit she punctures some of the oil cells in the rind and the oil kills the egg. Another interesting example is shown by the resistance of soya beans to attack by various insect pests in store. The resistance has been shown to be due to the presence of a chemical in the beans that interferes with the digestion of proteins by the insect.

Biological methods

By 'biological methods' is meant the use of one living organism to control another. One of the commonest of these is the use of animals to control weeds. Pigs, for example, are particularly adept at rooting out the underground stems of the bracken plant, and goats will eat almost anything. Geese are used in the weeding of mint fields in the northwestern part of the United States: they thrive on the weeds, but do not eat the mint. Ragwort, a plant which is poisonous to cattle, can be kept in check by the grazing of sheep. Grazing by cattle can reduce a number of weeds but

33

some plants have evolved a defence against such attack—the prickles of the thistle, the roughness of horsetail, the bitterness of buttercup—and the continuous grazing of a pasture gives these plants a good opportunity to thrive. The trampling of cattle can keep some weeds down and it is said that the removal of cattle from the Scottish hills and their replacement by sheep during the Clearances of 1782–1820 and 1840–54 led to the spread of bracken. Several species of fish have been observed to keep certain aquatic plants under control and fish have also been used in the control of mosquito and other larvae.

Biological methods may also be held to include the growth of competitive crops, i.e. crops that will successfully prevent the growth of weeds; thus subterranean clover has been used in Australia to smother the growth of St. John's wort. In Canada, two crops of buckwheat may be grown in one year and assist in the control of couchgrass and sow thistle. The farmer is himself a biological agent, and his various activities come under this heading. They are of course far too numerous to list in this small section but they are discussed under various headings throughout the book. It may be noted here however that by adjusting his soil conditions, the farmer can sometimes play a major role in the control of weeds, e.g. spurrey and sheep sorrel will flourish where there is a lack of lime; rushes thrive on wet waterlogged soils; and by adjusting these conditions the weeds can sometimes be eliminated without other methods. The main biological method is, however, known as 'biological control'. This consists essentially of the introduction of a living organism that will feed exclusively on the pest. Many examples abound and the topic is dealt with in Chapter 9.

Chemical methods

The development of chemicals for the control of pests has been one of the major agricultural and medical developments of this century. The pesticide industry has become one of the biggest sectors of chemical and biological activity in the Western world. The use of pesticides has enabled man to increase his agricultural yields, and the introduction of pesticides into the medical field, particularly to assist in the control of mosquitoes, fleas, lice, flies and other disease-bearing insects, has transformed man's environment, particularly in the tropics, and has helped to create a population explosion that might well promote the development of world planning of birth control, agricultural production, and resource utilization which may eventually lead to the dream of the centuries—world government. The development and use of chemicals for pest control forms the main bulk of this book.

Chapter 3

THE DEVELOPMENT
OF A PESTICIDE

This is not a particularly easy chapter to write. Each chemical company has its own methods of finding and developing pesticides and these methods differ to a greater or lesser degree from those of other companies. There is no one method; nevertheless, there are some common threads. This chapter has been compiled after reading papers and listening to talks by industrialists and others and taking part in discussions with them.

The pesticide appearing neatly-packaged on the shop shelf or in the farmer's shed is the end-product of many years of work. Its presence owes much to research, development, marketing and money. The finding of a chemical that is capable of killing a pest or pests is but one aspect of a long chain of events involving a number of major decisions: are the raw materials easily procurable ?, is the manufacturing process feasible ?, is it safe to use on food crops or in water ?, what is its effect on the environment ?, does the financial outlay justify the economic return ?

The financial outlay is certainly considerable—the spending of over £1m., and some six to eight years' work lies ahead after a promising chemical has been detected. But first find your chemical. There are three main methods of so doing.

Random or speculative screening
In this method almost every chemical synthesized by the company is put through screening tests, no matter what the original purpose for which it was intended (photographic, pharmaceutical, perfumery etc.), to find out if it shows any activity as a herbicide, insecticide, fungicide, rodenticide, molluscicide and

so on. If it does then it may itself be eventually marketed, or it may give the chemists a lead in synthesizing related chemicals and therefore possibly give rise to a new series, thus aminotriazole (see p. 70) was an intermediate in colour photographic dye production before it was developed as a herbicide; and the dithiocarbamate fungicides were first produced as accelerators in rubber processing. Occasionally a chemical may be found to have more than one exploitable activity, e.g. warfarin, which is used in medicine for thrombosis, is a very effective rat killer. It is obvious that the bigger the company and the more diverse its interests in the chemical field, then the more chance there is of success.

Directed screening
In this method the chemist studies the literature on compounds known to be biologically active and deduces from his reading that certain combinations of chemical groups are likely to show pesticidal activity. Once activity has been established in a particular group then the chemist produces related chemicals for testing. Thus the initial success of 2,4-D and MCPA as weedkillers (see Chapter 5) led to an investigation of the particular chemical structure causing herbicidal activity, and thus to a whole range not only of substituted phenoxyacetic acids but also many distantly related herbicides. The demonstration by the Geigy Company of the insecticidal power of DDT led to the development by many companies of insecticides of the organochlorine group. The advantage in this type of development goes to the companies with a major investment in the research chemical side.

Basic research
By studying the biochemistry of the pest it should be possible to formulate chemicals that will interfere with one or more of the basic biochemical activities. The development of the substituted phenoxyacetic and phenoxybutyric acids as herbicides (Chapter 5) stemmed from the investigation of the finding that indole-3-acetic acid was a fundamental plant hormone. The company with strong research interests, whether of its own or through support of pure research in institutes or universities, will be in an advantageous position. It should however be said that many of the major companies support research in universities and elsewhere without any thought or expectation of direct return; and indeed

D   37 no restriction is placed on the direction of research or on its publication. Companies may also place research contracts to have a particular piece of work carried out of direct interest to the company. In the opinion of the writer both activities have benefited university departments, not only in the support given but also by putting the academic in contact with practical problems and with his colleagues in industry. The contact has been mutually beneficial as each has learned to appreciate the work of the other. But in the pesticide field, it is a sobering thought that almost every breakthrough has come from the 'hit or miss' method. The return in terms of active compounds is small—the odds are about one in seven thousand five hundred against finding a winner—but with effective screening methods industry appears to find this a reasonable return. It is essential that a large number of compounds can be tested rapidly, and since the quantities of each chemical available in the first place may be small the test system must be devised in such a way that only small quantities of each are required. When a chemical does show activity (it has been calculated that on average 5 per cent do) then it is merely the beginning of a long process by which many chemicals showing initial activity will be eliminated for other reasons.

The screening methods vary from company to company but in general the chemical is tested (1) against a range of insects of economic importance, both by applying the chemical as a spray and by incorporating it into the insects' food; and (2) by applying the chemical to some 10–15 weed species growing in pots in the greenhouse (plants of economic importance may be included at this stage to determine selectivity). Both foliage application and soil application methods are used: (3) against a range of plant pathogenic fungi growing in test tubes in the laboratory and by application to a number of diseased plants in the greenhouse; and (4) against other pests ranging from eelworms to rats.

It is at this stage that the scientist carrying out the screening has to be particularly alert to spot some feature of the activity of the chemical that might be quite different from the one that he was expecting. For example, a chemical which was thought from its chemical structure to be a promising herbicide may be seen to be inducing root formation, early flowering, early seed production, stunting, increased shoot production and so forth—which might give a lead into a completely new use for the chemical.

The detailed procedure varies from company to company, but successful laboratory and greenhouse screening against a selection of pests is followed by testing in the field.

The company then takes the initial decision whether or not to proceed further, after preliminary tests have been carried out on rats and rabbits to determine mammalian toxicity, and after having established in a broad way the probable economic viability. Many chemicals (remembering that all showed initial activity) will be rejected at this stage. If the decision is taken to proceed, provisional patents will be taken out.

Mode of action studies are begun, together with information-gathering on residues, breakdown in soil etc. A good deal of far more detailed work is done on toxicity using rabbits, birds and fish and perhaps other vertebrates as well, by feeding known quantities of the pesticides to them and by applying pesticides to the skin of these animals. Long-term feeding studies with low doses, together with short-term feeding with high doses, are carried out. Various organs of the test animals are examined by eye, by analytical instruments and under the microscope. The metabolism of the pesticide within the mammalian body is also investigated. If these mode of action studies are satisfactory, widescale field conditions are initiated. Before this phase can be begun, however, large quantities of the chemical have to be made; and it has to be formulated in one of the conventional ways—in the form of dust, granules, water-dispersable powder, emulsifiable liquid and so on. Also during the second phase, while the field tests and toxicology studies are being carried out, a third activity is in progress: the manufacturing development process —determining the availability of raw materials for the manufacture of the pesticide and the best methods of synthesizing the pesticide. A more realistic cost estimate is made, the marketing prospects of the pesticide are assessed and the patents specification is completed.

At the end of this period—which may take one to two years— the second decision point is reached. If it is decided that work should proceed, then a pilot plant to manufacture enough of the chemical for world-wide testing on larger plots than the original tests is set up. Arrangements are made for treated crops and soils to be available for examination for residues. The question of residues is one of the most important the company has to face. All governments require information on pesticide residues on treated crops, which they consider in relation to the toxicity of the chemical, the breakdown products and other factors before sanctioning the use of the chemical on a particular crop. The crop which has been sprayed with the pesticide is then examined at intervals to determine the amount of pesticide that has been absorbed and still remains within the tissues. During these examinations very sophisticated equipment is used which enables the investigator to detect minute amounts of the pesticide.

There are some who would say that perhaps the methods are too sensitive; that minute amounts formerly unnoticed are now detectable at such concentrations that they couldn't possibly do any harm. For example the reason that we know that Adelie penguins at the South Pole have 0.18 ppm of DDT in them is because the technologist has developed such sensitive equipment. Be that as it may, limits for residues are set by many governments and if these limits are exceeded in tests then the pesticide, whatever its other merits, is rejected by the company. Bioassay methods are sometimes used to detect small quantities, particularly where the chemical is a new one and instrument-based methods may not yet have been worked out. Bioassay involves the use of an organism that is particularly sensitive to the chemical; for example one millionth part of a gram of the herbicide 2,3,6-trichlorobenzoic acid can be detected by the sensitivity of french bean seedlings to it.

Houseflies are among the most easily available insects and are commonly used in the bioassay of insecticide residues. Using feeding methods, 0.1 ppm of dieldrin or aldrin, 0.18 ppm of gamma-BHC, 0.5 to 0.6 ppm of endrin and 1.4 ppm of DDT can be detected in milk by young flies and micro-quantities of DDT, gamma-BHC and parathion can be detected in chicken flesh, cabbage, apples, cucumbers, strawberries and peas. This method may also be used in field laboratories where sophisticated equipment is not readily available.

The toxicology studies are then stepped up and carried out in greater detail. Long-term exposure tests with the chemical and with its metabolites are carried out on rats and dogs. Some of these exposures may be life-long. Examinations are also made of the rate and quality of the reproduction of animals fed on, or exposed to, the chemical, including both residue and toxicological studies. The cost of providing the necessary evidence of safety may be anything from £10,000 to £500,000. The manufacturing process is examined in detail and this may involve the construction of a plant or the 'farming out' of the process to a company which already possesses the right type of plant to cope with the process. The economics of the product are examined in considerable detail, including the proposed market for the product. At the end of this evaluation phase a third decision point is reached. If all the tests have proved to be satisfactory further development work will be carried out. All of the toxicological and residue information that has been accumulated is made available to the appropriate authorities in the various countries where the compound is likely to be used. These authorities may state that more information is required; if so then further studies are carried out by the company. At the same time data sheets and instructions for use are prepared and worldwide trials, often in co-operation with proposed marketing organizations and companies are put under way. Packaging materials are devised and constructed.

At the end of this third phase, which will probably take two years to complete, the fourth and final decision is taken whether or not the company should go into full commercial manufacture. All the various data are assessed—effectiveness as a pesticide, safety for the user, safety for wildlife, availability of raw materials, cost of manufacture, economics of the pest to be controlled and so on. In addition the results of co-operative work that may have been proceeding with research institutes, universities, and colleges on the mode of action of the pesticide, metabolic studies, the effects on the microbiological population of the soil, field trials, the effect of the chemical on the 'ecological balance' of the countryside (does it injure predators as well as pests ?, does its action interfere fundamentally with an important food supply of a desirable animal ? etc.) are taken into account. A decision to proceed involves the erection of a new plant which in a further two years' time will be producing the finished material in commercial quantities. In Britain the chemical and biological data will be presented to the Pesticides Safety Precautions Scheme. Such submissions are admittedly carried out on a purely voluntary basis, but it should be noted that no reputable firm omits this part of the procedure. The Pesticides Safety Precautions Scheme is administered by the appropriate department of the government, acting in full association with other government departments and calling on other expert committees, e.g. the Advisory

Committee on Pesticides and other Toxic Chemicals, for advice. Under this scheme no pesticide product is put on the market until conditions have been agreed which ensure that there shall be no hazard to users or consumers of treated foodstuffs and that risks to wildlife are reduced to a minimum. The onus is on the company to provide all the evidence necessary to support adequately the claims regarding the safe use of the product. If the product is given clearance then the company is informed. Official recommendations are then published in a loose-leaf booklet en-entitled 'Chemical Compounds used in Agriculture and Food Storage in Great Britain. User and Consumer Safety—Advice of Government Departments'.

The advice is usually given in three parts—operator safety, consumer safety, and safety to others including wildlife. The government has the right to review existing recommendations on a pesticide at any time in the light of more recent knowledge, and if necessary, to revise the recommendations.

The Agricultural Chemicals Approval Scheme is concerned with the performance of the product, ensuring that it measures up to all the claims that are made for it. Ministry officials have access to trial plots and all the records that the company possesses relating to the proposed pesticide; and they may, if they so wish, arrange for independently conducted trials to be carried out. Approval, if eventually given, entitles the manufacturer to use the official approval mark on his labels. Approval will not of course be given unless the product has also passed the Pesticides Safety Precautions Scheme.

The requirements to be met regarding toxicity and efficiency vary from country to country, and these requirements must also be met by the company. In the United States, for example, 'petitions for tolerances' and 'granting of labels' operate at both state and federal levels.

It can be seen from the above that no company would enter lightly into the production of a pesticide. The cost is high and the sieve is a fine one. The popular picture that is sometimes portrayed of the pesticide manufacturer encouraging wholesale chemical spraying over the countryside is a completely false one. Apart from the moral considerations which he shares with other members of the community he cannot afford to have his pesticide do harm and thus be withdrawn: his financial commitment is far too heavy. For the first four years that a new pesticide is on the market, the company will be losing money: it is only by the fifth year that some profitability begins to show. And most pesticides have a market life of no more than ten years.

There is another major risk involved as well: since it takes 6–8 years to put a pesticide on the market, the need for it which was very apparent when the development was begun may have disappeared before the development is completed—perhaps owing to the pest decreasing in importance, or a control other than the chemical one being evolved, or quite simply because another chemical has come first onto the market. This is a calculated risk that the manufacturer must take and he must have plenty of capital behind him in order to take it.

One of the criticisms that has been directed against the pesticide company is that it is interested only in large-scale problems —problems big enough to yield an adequate financial return for the enormous outlay that is involved. In general this is true: a company has to make profits in order to survive. What, then, of the small local problems that are perhaps crying out for solution but are so small that they are not worth the investment? Some may be solved as side issues to the major developments.

It has long been recognized that bracken is a problem in Scotland. The plant covers about half a million acres of hill-land and is spreading. The land is not, however, of sufficiently high value for a chemical company to consider developing a herbicide specifically for bracken. But now it looks as if asulam, a herbicide developed by May & Baker Ltd for docks, will also be a very effective bracken killer.

Another complication arises in that the cost of a herbicide may be more than the land justifies. In such instances there may be a case for a government subsidy such as there is at present for bracken cutting. It may well be that this is the pattern that is required—co-operation between government and industry for the solving of such problems—not only to subsidize the use of a particular herbicide but even to subsidize its development.

Another criticism sometimes levelled is that the screening process is a very wasteful one, since similar compounds may be screened over and over again by different companies. Most industrialists would agree that there is waste, but would probably aver that the waste is more than offset by the advantages of competition.

As has been noted the great majority of new pesticides derive

43

from research conducted by the chemical companies, and the marketing of existing pesticides over which companies possess patent rights is a major means of financing such research. It has been suggested further that the development of pesticides by companies leads to a multiplicity of pesticides of similar properties. In practice it is rare for two pesticides from different sources to be so similar that a worthwhile difference in use does not occur in some part of the world.

Industry must constantly be looking ahead not only for new chemicals but also for new developments in pest problems. The pest problems of today will not be the pest problems of 15 or 20 years hence. Predictions are notoriously difficult things to make but trends can often be discerned and for the discerning the rewards may be high. It was predictable that the continued use of 2,4-D or MCPA herbicides would lead in time to the control of most broad-leaved weeds. How many would have predicted that this would lead to the dominance of grass weeds resistant to 2,4-D and MCPA? And of those who did predict, how many had the capacity to start developing herbicides specifically directed at the grasses? How many predicted that the use of insecticides would increase the importance of the resistant mites? Or that the malaria-carrying mosquito would soon develop resistance to DDT? It is easy to be wise in hindsight. It takes real ability to look ahead.

Chapter 4

INSECTICIDES

The use of chemicals for insect control goes back a long way. Around 1000 B.C., Homer referred to a pest-averting sulphur, and about 200 B.C. Cato advised boiling a mixture of bitumen so that the fumes would blow through grape leaves and rid them of insects. In the medical field, hellebore was used by the Romans for the attempted control of lice. The Chinese were using arsenic to control garden pests before A.D. 900 and Marco Polo writes in 1300 of the use of oil for camel mange. By the second half of the nineteenth century substances were being used some of which are still in use today—arsenical compounds were proving reliable against Colorado beetle and gypsy moth in the USA, and nicotine which had been prepared for years as a simple infusion of tobacco leaves effectively controlled such insects as aphids, thrips and mites.

The effectiveness of the tobacco infusion spray is due to the presence of the alkaloid nicotine, which is present in tobacco leaves in a concentration of up to 14 per cent. Nicotine is nowadays obtained commercially from tobacco by steam distillation or extraction by solvents. It can also be synthesized in the laboratory but the natural product is cheaper and is therefore preferred. It is used as a contact insecticide for aphids attacking fruits, vegetables and ornamental plants; and as a fumigant in greenhouses and against poultry mites.

Although man has been engaged in a constant struggle with pests throughout the ages, the battle has been woefully one-sided. He has had pitifully few weapons at his command. One of the most potent, however, was pyrethrum.

In the early 1800s a Mr. Jumtikoff, an Armenian, discovered that certain Caucasian tribes used ground flowers of a plant as a flea and louse powder. In 1828 his son began making and selling this product to which he gave the name 'pyrethrum', by which it is known today. It is got by extracting the ground flowers of various species of chrysanthemum (the old name for the genus is *Pyrethrum*) with a solvent such as acetone. The mixture is then passed over charcoal to get rid of impurities. It has been found that the active agent is located mainly in the little 'seeds' (*achenes*) of the flowers. It is probably the oldest of the organic insecticides, and interestingly enough although it is more than 145 years since it was first introduced it is still going strong and has lately displaced some of its new competitors, particularly for domestic use. Synthetic pyrethrums have also been developed.

Despite much investigation the biochemical mode of action of pyrethrum is not known. It readily penetrates the insect cuticle and paralyses the insect. There is generally a rapid 'knock-down' followed by a substantial recovery so that an approximately three-fold increase in dosage is required to ensure that a knocked down fly is a dead fly. This recovery from paralysis is due to the insect having the ability to detoxify the poison by enzyme action. It is possible to add another chemical to the pyrethrum which will prevent such enzyme action by the insect, and both piperonyl butoxide and sulphoxide have been used for this purpose. They have considerable importance in 'aerosol bombs' and household and cattle sprays. Pyrethrum is of low mammalian toxicity.

For centuries primitive people in many countries have used 'doctored' baits of all kinds to stun fish. In 1848 a Mr. T. Oxley suggested that a Malayan fish poison extracted from the root of a leguminous plant *Derris eloptica* could be a good insecticide. This was shown to be the case and derris dust was accepted as a potent insect killer. The active principle, rotenone, has since been shown to be present in the roots of other plants such as *Lonchocarpus* from South America and *Tephrosia* from East Africa.

The first synthetic insecticides were inorganic. In 1867 the pigment Paris green, a crystal compound of acetate and arsenite of copper, was used successfully in the USA against the Colorado beetle of the potato fields and later against a great variety of leaf-eating insects.

Sodium fluoride has, since the 1840s, been known to be highly toxic to insects; it is used mainly as a barrier against crawling insects such as cockroaches, earwigs, and ants. Fluorides may be incorporated into roof timbers to protect them against termite attack. Boric acid is still widely used against cockroaches, being dusted into their hiding places. It is very effective in these situations. The dinitrophenols have a wide range of biological activity, for not only are they selective weedkillers (see Chapter 5) and fungicides (see Chapter 6) but they are also potent insecticides.

Potassium dinitro-2-cresylate, marketed in Germany in 1892, was the first synthetic organic insecticide. This led to the development of related chemicals, some of which are used as sprays for killing the eggs of mites, aphids and scale insects. Others are used as foliage sprays and DNOC in oil is used to control grasshoppers.

One of the first synthetic organic insecticides to be widely used was 'Lethane' (an organothiocyanate compound), which was marketed in 1932 as a rapid knock-down agent for the control of household insects. The members of this group of insecticides have a rapid paralytic action, and are used principally as household aerosols and sprays for the control of flies, mosquitoes, cockroaches and as dairy and livestock sprays for fly control. Their rapid knock-down makes them very useful for attacks on flying insects but some of them have rather unpleasant odours.

It was, however, the Second World War which provided the greatest stimulus for the development of organic insecticides manufactured in the laboratory. This was brought about not only because of the need for increased food production with a smaller labour force but also because armies were moving into mosquito- and fly-ridden regions and it was necessary to control these and other insects. Interestingly enough, although the powers of DDT as an insecticide were not recognized until the early 1940s it was synthesized more than 100 years ago.

ORGANOCHLORINE INSECTICIDES

In 1874 a German chemist, Dr. O. Zeidler, synthesized a chemical with the rather daunting name 1,1,1,trichloro-2,2,di-(4-chlorophenyl) ethane. It was a white powder of no apparent use. It was put into a bottle, labelled, and put on the laboratory shelf— where it was to lie practically untouched for 65 years (how fortunate that it was so stable!) until the demands of war in 1939 brought it to the fore, not as a weapon of war, but as a means of saving lives.

Its insecticidal powers were discovered by a Dr. Paul Müller, working for J.R. Geigy in Switzerland. It was found to be effective against the Colorado beetle, fleas, lice, mosquitoes, flies and many other insects. Samples were sent secretly to Britain and to the United States of America where further tests were carried out. Volunteers went for days in underclothes treated with this chemical, now known as DDT, and three weeks after treatment clothes were still lethal to lice. No volunteer showed any signs of poisoning. It was used for treating soldiers' uniforms, and it is of interest to note that at that time the rumour went round that it had an adverse effect on male virility. This was completely unfounded: the birth rate shot up at a staggering rate after the war and has been going up ever since. One of DDT's triumphs was in 1944 when it controlled a major outbreak of typhus in Naples.

And so the first of the organochlorines was born. Despite the fact that DDT was one of the first organic insecticides to be used and is still the most widely used insecticide, it is only recently due to the work of Drs. Matsumura and O'Brien that we have come to have an understanding of how it works. Apparently, it complexes with the energy-producing enzyme system at nerve endings, causing the characteristic tremors and convulsions that are well-known features of DDT poisoning. The site of action is established as the peripheral nervous system, since it has been shown that the isolated central nervous system is insensitive to DDT.

Following on the discovery of the insecticidal action of DDT, other organochlorine compounds were examined and synthesized. One of these, benzene hexachloride, was first prepared by Faraday in 1825; but its insecticidal properties were not discovered till about 1940 when British (I.C.I.) and French workers independently developed it. A purified form, gamma benzene hexachloride or gamma-BHC, is sometimes known as lindane after Van der Linden (who first isolated it without knowledge of its insecticidal powers in 1912). It has a broadly similar action to DDT. Both compounds can induce resistance in insects (see p. 56) but resistance to gamma-BHC is more likely to be coupled with resistance to the cyclodienes (see below) than to DDT. The taint caused by crude BHC, which was evident even at very

48

low concentrations, has led to its being condemned for use on food crops and although the purified form gamma-BHC has little taint the prejudice remains. It is used as a soil pesticide, as a seed treatment, for grasshopper control and against cotton insects. Its mode of action is unknown, though it is thought that it may be similar in action to DDT. It is one of the less persistent of the organochlorine compounds and has only a moderate toxicity to mammals. Related to gamma-BHC is toxaphene which is a mixture of several compounds which have always been formulated without separation. It has similar biological properties to gamma-BHC.

Cyclodienes
The insecticidal properties of chlordane were first described in 1945 by Kearns and co-workers, and this was the first of a series of compounds which were made using the so-called 'diene' synthesis. This reaction was discovered by two chemists, Diels and Alder, and for this reason the most important members of this group of insecticides were given the names dieldrin and aldrin.

The starting-point for the production of chlordane is chlordene, which was discovered by a Dr. Hyman in 1945. Chlordane contains heptachlor as an impurity, and this has been found to be a more effective insecticide than chlordane itself. Other variations of the Diels-Alder reaction produce aldrin, which can in turn be oxidized with hydrogen peroxide to yield yet another powerful insecticide—dieldrin. Dieldrin is also formed from aldrin in living tissue; another close chemical relative is endrin.

Chlordane and heptachlor are used for the control of cockroaches, ants, termites and other household pests. Aldrin, dieldrin and endrin are the most powerful general insecticides known. Aldrin is volatile, and this makes it particularly useful for soil application; but it is also used as a broad spectrum insecticide for the control of insect pests of fruits and vegetables, and as a seed treatment. Dieldrin is particularly active against the parasites found on the skins of sheep and cattle and it has been used for this purpose as a dip against lice, ticks, and blowflies. It is also used for the mothproofing of materials. Endrin is very useful for the control of caterpillars attacking field and vegetable crops; and it is one of the few pesticides which is effective against certain species of mite otherwise difficult to control.

49

Insecticides

The symptoms of poisoning by all of these cyclodiene insecticides are similar. Affected insects show convulsive activity with final complete prostration. The insecticides are readily absorbed through the cuticle of the insect and the site of action appears to be the central nervous system rather than the peripheral nervous system which is affected by DDT. Treated insects show a characteristic 'fanning' of the wings. There are interesting variations of toxicity within the group; thus endrin is relatively non-toxic to the housefly whereas aldrin is highly toxic. The biochemical mode of action is not known. Most of these compounds, especially aldrin, dieldrin, and endrin, are very stable and resistant to breakdown and this has led to the introduction of other cyclodienes which are less persistent. Endosulfan, for example, has been produced by the Hoechst Chemical Company in Germany and telodrin by Shell in Britain.

ORGANOPHOSPHORUS INSECTICIDES

The first indications that organophosphorus chemicals were toxic came not from insecticides but from bootleg liquor in the United States of America during the prohibition period. A chemical called tricresylphosphate was manufactured originally as a plasticizer but was also used occasionally as an adulterant of cooking fat and more widely as a ginger flavouring for some rather unpalatable and questionable 'whiskeys'. Some of the illegal liquors were no doubt poisonous enough in themselves but the addition of tricresylphosphate made even the most innocuous liquids deadly. Many deaths and many cases of paralysis of the limbs resulted.

Research into the possible use of organophosphorus compounds as war gases was begun during the Second World War in the laboratories of Britain, France, Germany, and the USA. The research was very successful, but the products were fortunately never used. Among them were some of the most potent nerve gases known to man. But among them too were some which, while not so toxic to man, were very potent against insects. And so a whole new group—the organophosphorus insecticides—was born. In 1943 Dr. Gerhard Schrader of the Bayer Company's laboratories in Germany produced the first organophosphorus compound available for crop protection—parathion. There
50

followed in rapid succession schradan and demeton (the first practical systemic insecticide; see p. 52), azinophos-methyl (a broad spectrum insecticide), and trichlorphon, a stomach poison.

Many of the early organophosphorus compounds including parathion (trichlorophor is an exception) are very toxic to mammals, including man. Research has therefore been directed towards finding compounds that are highly toxic to insects but which are of low mammalian toxicity. Such a chemical is malathion which was introduced in 1950 by the American Cyanamid Company. Although only about a half to a twentieth as toxic to insects as parathion, it is only about one two hundredth parts as poisonous as parathion to mammals. Malathion, which is an outstandingly successful insecticide, and is probably the most widely used insecticidal organophosphorus compound, has a low mammalian toxicity because it is rapidly broken down in the livers of animals. In the insect, on the other hand, it is very stable and can therefore exert its toxic effect: it is widely used in the tropics for mosquito control. At the other end of the scale it is also widely used by amateur gardeners for the control of pests. Its world-wide use in million pound quantities, in the home and in the garden, in the orchard, on animals and in insect eradication campaigns indicates that it is one of the safest of insecticides.

It is estimated that more than 100,000 different organophosphorus compounds have been synthesized and tested out as insecticides. At present more than 40 are commercially successful, a number of them being manufactured in multimillion pound quantities.

Some of the members of this group are contact poisons whilst others are systemic (see below). The fact that the group contains many chemicals which are potential war gases has led to their mode of action being more thoroughly investigated than any other group of pesticides. Their main and primary action in mammals is as powerful inhibitors of a group of enzymes known as the cholinesterases; and it seems to be clear that a similar action takes place in insects, though the cholinesterases in the two groups—mammals and insects—are not identical.

Schradan, parathion and malathion are weak anti-cholinesterases themselves but are apparently converted in the insect to powerful inhibitors by certain enzyme systems.

The chemical acetylcholine is the transmitting agent of nerve impulses in insects and in mammals. Although it serves a very necessary purpose it must not accumulate indefinitely at the nerve junctions, or there is a gross interference with the co-ordination of muscular response. Such interference with the muscles of the vital organs produces serious symptoms and eventual death. It appears that the organophosphorus compounds inhibit the activity of the enzyme acetylcholinesterase which is normally responsible for the removal of accumulated acetylcholine. This leads to a build-up of acetylcholine in the nervous system of the organism, with the results indicated above.

Systemic insecticides

Systemic insecticides are chemicals which are absorbed by the plant and translocated to its various parts in amounts lethal to insects feeding thereon. They may enter the plant by direct absorption through seeds, roots, stems, leaves, or fruits and all of these methods have been used not only experimentally but also on a practical scale.

This method of protecting plants against attack by insects has a number of decided advantages over the normal spray methods. These are: (1) there does not have to be complete coverage of the plant; (2) the chemical is protected within the plant from the effects of weathering; (3) new plant-growth taking place after application will also be protected, since the systemic insecticide will move into these new parts as they are formed; and (4) there is much less risk of damage to harmless and beneficial organisms.

In order to be completely effective the chemical must have: (1) sufficient water solubility to enable it to move in the plant sap; (2) the ability to penetrate into the plant through roots, stems and leaves; and (3) sufficient stability in the plant environment to enable it to exert the required degree of residual insecticidal action. In order to make the systemic compound acceptable to public health requirements, a further property must be added: (4) the chemical must be subject to decomposition in the plant tissue so that it breaks down to non-toxic products within a reasonable time.

There are now a considerable number of insecticides which act systemically. These are mainly organophosphorus compounds; and they include dimefox, demeton, demeton-S-methyl, dimethoate, mevinphos, phosphamidon, and dicrotophos. Systemics have caused great advances in the control of inaccessible

52

pests, such as certain aphids secreted in crinkled leaves or even galls, leaf eelworms secreted in leaf tissues, red spiders under bud scales, aphids and mealybugs hidden away in cracks, or even insects protected from the spray simply by sitting on the underside of the leaf. Mevinphos gives excellent toxicity of a very transient nature, and is particularly useful on crops harvested shortly after treatment, such as lettuce. Demeton is used commercially in the USA, Europe and parts of Africa against aphids on apples and pears, potatoes, strawberries and musk melon. It is sprayed by aeroplane, and it has been widely used in the USA and the Sudan to protect the cotton crop against cotton aphids. Demeton-S-methyl is also employed in Europe against the same range of pests as demeton. One of its most important uses in Britain has been for the protection of sugar beet against attack by those aphids which carry the virus causing the disease known as sugar beet yellows, which can be disastrous to the crop. Systemic insecticides in general have shown a great deal of promise against soft scales and mealybugs, leaf hoppers, leaf miners, the olive fly, leaf-eating insects, mites such as the red spider, and foliar eelworms.

Some systemic insecticides such as phosdrin act directly on the cholinesterase enzyme system; others however are non-toxic in themselves, and only become insecticidal when they are converted into a toxic substance within the insect.

Systemic organophosphorus insecticides have found a use not only in plants but also in the protection of animals against pests. They can either be sprayed on to the animal, in which case they are absorbed through the skin, or they can be incorporated into feeding stuffs. They have been used against internal parasites such as cattle grubs, screw-worm larvae, and worms of all kinds; and against external parasites such as stable flies, mites, lice and ticks—in all cases without injury to the host animal. The insecticides are slowly destroyed by enzymatic activity in the animal body so that after a safe period the animal can be used for milk production or slaughtered for meat. Among the insecticides used are fenchlorphos, which is fed to the animal, coumaphos, and crotoxyphos which are used as sprays, and cruformate which can be either fed or sprayed.

The organophosphorus compounds represent the most versatile group of insecticides yet developed by man. Some of them such as diazinon have fairly long residual action; others such as

E

mevinphos rather have a very short residual action. Some such as parathion and malathion will kill a whole range of insects, while others such as schradan are very specific and only toxic when converted by aphids, mites and so forth. Some can be used as systemic insecticides in plants, others as systemic insecticides in animals. Some have rapid contact action; others such as trichlorphon have no contact activity but are effective stomach poisons. Some such as parathion are very poisonous to mammals, whereas others such as malathion are virtually non-toxic to mammals.

CARBAMATE INSECTICIDES

Although these insecticides are quite different chemically from the organophosphorus insecticides they have a killing action which is very similar to the latter both in symptoms and in detail; they attack the cholinesterase enzymes and cause poisoned insects to exhibit violent convulsions and other neuromuscular disturbances. They penetrate rapidly into insects, although some insects are able to detoxify them, and are therefore resistant. This detoxification can be prevented by adding other chemicals, such as piperonyl butoxide, to the carbamate. The carbamate insecticides were introduced by J.R. Geigy.

A wide range of such insecticides is now produced by many companies, some being highly specific. Dimetelan is used as bait with sugar or impregnated bands for the control of houseflies. Carbaryl is the most widely-used carbamate, being employed on a variety of crops such as cotton, fruits and vegetables. Aminocarb is active against many pests including snails and slugs; propoxur is used in houses to control flies, mosquitoes, ants and cockroaches. Methiocarb is used against fruit pests, and aldicarb is a soil-applied systemic.

SELECTIVITY OF INSECTICIDES

It is not enough that an insecticide should kill insects. It must also have a degree of selectivity, so that it is more toxic to insects than it is to plants or to mammals, including man; it should in addition preferably be more toxic to harmful insects than it is to beneficial ones. The former is more easily achieved than the latter but both are desirable. Selective insecticides are required for two reasons: (1) to avoid unfavourable changes in the environment; and (2) to reduce hazards involved in their use. There are various ways of attaining selectivity, which are dealt with in detail by Dr. R.D. O'Brien (see further reading).

Care in application is important—directing the spray so that the insects are covered by the poison while the operator is not. There are numerous devices such as protective clothing, respirators and so on to aid this type of selectivity from the operator's point of view. The protection of beneficial or harmless insects is not so easily attained although the timing of the application and the use of systemic insecticides help to achieve this. When different animals are covered by the same spray, then selectivity is physiological; the following are examples.

Differential absorption

Some animals, although covered by the same amount of spray, will absorb insecticide more readily than others. The three main routes of absorption are via the skin, the digestive system and the respiratory system. The skin of mammals and the integument of insects are such vastly different tissues that one would expect that some chemicals would penetrate one and not the other. Furthermore all integuments are not the same, so we find that the integument of the Japanese beetle and the American cockroach readily allow the passage of DDT whereas those of the milkweed bug and the bee present a barrier to it. Most chemicals are more poisonous when taken by mouth than when applied to the skin, but again we note selectivity: the oral toxicity of DDT to mammals is low but to the bee it is very toxic by this route. Indeed, O'Brien considers that it is possible that absorption selectivity may account for the low mammalian hazard of many chlorinated hydrocarbons.

Differences in metabolism

All the organic insecticides can be broken down in the bodies of animals. It has now been shown that some insecticides such as malathion are less toxic to humans, because we have the ability to break them down rapidly in our livers. Insects do not possess this ability and are therefore poisoned. In addition, certain insects may possess the power to break down some insecticides while others do not. The former insects are therefore susceptible to the insecticide while the latter are resistant.

Differences in disposal

Insecticides can be disposed of either by excretion or by storage. The excretory systems in insects and mammals are so very different that one would expect differential excretion, but this is not so common as one might suppose. However, storage is of great importance with certain insecticides—DDT is extensively stored, and thus immobilized, in the fat of mammals. It is thought that certain insects may be more resistant than others because of their ability to store DDT in fat bodies.

Differences in penetration to the target area

In order to exert its toxic effect the insecticide has to get to its target site, i.e. the organophosphorus compounds have to penetrate the central nervous system. It is thought that some insects may be resistant to attack by certain organophosphorus compounds because they have a protective sheath around the nerve which prevents penetration by these compounds. O'Brien considers that a close study of these various factors (and others) should enable selective toxicity to be built into a toxic compound. This would open up wide horizons in the pest control field.

RESISTANCE TO INSECTICIDES

The discovery of the high insecticidal action of DDT in 1940 was followed, as we have noted, by the development of related chemicals. This, plus the flow of organophosphorus insecticides coming out of Germany in the 1940s, and the demonstration by J.R. Geigy that certain carbamate compounds were potent insecticides, raised high hopes that man's insect problems were all but solved. But time was to unfold a very different story.

One of the major biological lessons that we learn is that organisms survive as a species because of the great variability within the species. No matter how the external conditions may change there are generally a number of individuals within the species which are able to tolerate the changed conditions, even if the rest of the individuals are killed off by the change. This phenomenon holds good throughout the world of living things—whether they be viruses, bacteria, fungi, insects, fish, birds or mammals (including man). Great plagues, for example, have decimated mankind but they have never wiped out the species Man. Some humans had a natural resistance to the plague, and they survived and bred. The same holds true for insects. Although DDT killed off most of the anopheline mosquitoes which were exposed to it, some were resistant, and they survived and bred. Today more than 200 species of insects are known to be resistant to one or more insecticide. More than half of this number are DDT resistant and more than a third are resistant to the other organochlorines such as gamma-BHC, dieldrin, endrin, toxaphene or chlordane. One of the most worrying aspects is the resistance of the malaria-carrying mosquito, *Anopheles*, to DDT. Another is that rat fleas capable of carrying plague are showing similar resistance.

It appears that resistance is not developed during the lifetime of an insect; for example, houseflies exposed daily to sub-lethal doses of DDT became progressively more susceptible to DDT, not progressively more resistant. Fruit flies were reared for 50 generations in a sub-lethal concentration of DDT without becoming less susceptible. It would appear then that the presence of pesticides does not cause insects to become resistant to them. Instead the indications are that resistant strains exist before the introduction of an insecticide; when the susceptible strains are killed off, the resistant strains then multiply to take their place. Resistance is heritable so that it is passed on from generation to generation: insects descended from resistant insects are themselves resistant. The following points have been made by Dr. Winteringham, Chairman of the Working Party set up in 1965 by the Food and Agricultural Organization of the United Nations 'to advise (the Director General) on all matters relating to pest resistance to pesticides'.

(a) The incidence of resistant strains is increasing wherever chemicals are systematically used for insect control. The present figure of 200 species may be compared with the figure of 60 or 70 agricultural pests reported in 1961 and is a sobering indication of the spread and ubiquity of this threat to chemical control, (b) some level of resistance almost invariably extends to related pesticides to which the insects have never been exposed, (c) there is accumulating evidence that the genetic potential for a significant degree of resistance to all the principal groups of available insecticides exists in most and probably all insect orders and species.

Dr. Winteringham reached the following conclusions:

(1) Increasing world demands for more efficient crop production and protection against pest attack seems likely to intensify the needs for safe and efficient pesticides for some years to come; (2) The problems of resistance have become more intense and widespread during the last decade and the trend may be expected to continue where chemicals are systematically in use for pest control; (3) The onset of resistance may be delayed and possibly obviated by avoiding all forms of unnecessary selection pressure e.g. by avoiding the use of unnecessarily persistent chemicals, and by the chronological and geographical restriction of chemical application to that necessary to keep the pest population at economic injury levels; (4) It is very important to detect and characterize resistance as soon as possible. This can be achieved by continuing surveys of pest populations under systematic chemical control. Test methods should be standardized to facilitate valid comparison over a period of time; (5) The evaluation of alternative chemicals should be undertaken in anticipation of resistance where chemicals are in systematic use and taking into account known cross resistance phenomena and pesticidal history of the area; (6) There is an urgent need for a wider range of pesticides of novel modes of action or requiring novel biochemical detoxication mechanisms; (7) There is a need for a much better appreciation and study of the overall biology of the pest so that chemical control may be directed with a view to preventing resistance in the first place.

At present the only method at our command is to switch to a new insecticide when an insect shows resistance. Sometimes, however, the chemical structure of some insecticides is so similar that resistance to one almost automatically means resistance to the others. But this is far from being universally the case, e.g. resistance of the rice-stem borer to parathion does not extend to parathion-methyl. Generally an insect resistant to one chemical will remain susceptible to others with fundamentally different chemical structures: the switch from organophosphorus compounds or vice versa is therefore possible. Another hopeful sign is that a resistant population, say of houseflies to DDT, will become susceptible if DDT is withdrawn for a year or two. This may be a general phenomenon although there is of course the ever-present danger that resistance may develop more rapidly than before, when the chemical is re-introduced.

PESTICIDES FOR ORGANISMS RELATED TO INSECTS

As mentioned in Chapter 1 there are a number of pests which, though not insects, generally come into the sphere of the entomologist. Among them are (a) mites and ticks, (b) slugs and snails and (c) eelworms.

Acaricides
This is the general name given to pesticides used against mites and ticks, and a number of pesticides have been developed specifically for this purpose. In general these chemicals are inactive against insects but attack mites and ticks either at the egg stage—the so-called ovicides—the nymph stage, the adult stage, or all three. Some of them are very specific. Among the pesticides that are active against all stages are chlorfenethol and dicofol. DCPM is fairly specific, being active against citrus and European red mites. Among the ovicides the best-known are chlorbenzide, chlorfenson and azobenzene, an orange dyestuff effective against the eggs of spider mites as a spray or when volatilized by burning in a mixture.

Molluscicides
A wide variety of chemicals have been developed for killing slugs and snails. There are two main groups, those that are used against land-living species and those that are active against water-living species.
The most effective chemical against land snails and slugs is

59

metaldehyde. This chemical was first developed as a domestic fuel for cooking, and it is said that its snail-killing powers were first observed when some was spilled at a picnic party. It is a toxicant for snails and is used as a bait mixed generally with bran. As is well known, slugs and snails exude a trail of slime. Metaldehyde which is absorbed by the foot of the snail as well as orally causes increased secretion of slime and thus loss of water and it seems possible that the metaldehyde acts largely as a desiccating agent. A number of carbamate insecticides including 'Isolan', 'Zectran' and methiocarb have also been found to be very active against snails when used in baits; and DNOC has been observed to reduce slug damage when used as a selective herbicide in a cereal crop. Triphenomorph has been developed by Shell Research Ltd as a pesticide active against aquatic and semi-aquatic snails. Details of a new drug have recently been announced: diamphenethide, which offers a resh approach to the treatment of liver fluke in sheep. As we have noted, several water snails harbour the bilharzia fluke. Copper sulphate at very low concentrations (2–3 ppm) will kill all water snails within 24 hours, but although it is cheap there is the possibility of damage to fish and other aquatic life. Pentachlorophenol, which is also very active, has the same drawback. Niclosamide is a water-snail toxicant introduced by the Bayer Company in Germany. It is very effective and will kill snails within 24 hours of exposure to a concentration in water of under 1 ppm. One of the most effective molluscicides used today, particularly against the vectors of bilharzia, is probably trityl morpholine.

Nematicides

Eelworms belong to the group of animals known as nematodes, and pesticides developed against them are called nematicides. Various eelworms attack the roots of certain crop plants and do immense damage. Chemical control has been most successful, through the use of soil fumigants which are sufficiently volatile to penetrate through the upper layers of the soil. In order to achieve good eelworm control there must be good distribution of the chemical in the soil. Many of the fumigants that are used are also active against insects, weed seeds and fungi in the soil. For general fumigation, for example of grain in containers, the compounds generally used are those that boil at room temperature such as hydrogen cyanide, methyl bromide and ethylene oxide.

In soil fumigation, on the other hand, exposure is much longer; and the slow releasing of gas from certain substances—for example, dichloropropene or ethylene dibromide—is particularly effective.

The first soil fumigant to be introduced was carbon disulphide, which was first used in France in 1858 in order to control pests of stored grain, and later in Germany in 1871 against the sugar beet eelworm. Chloropicrin was first tested in Britain in 1919 and was used extensively in the United States. In 1943 DD was introduced by the Shell Development Company; this began the era of commerical soil fumigation. The mixture is effective in killing eelworms and their cysts, but is toxic to other life-forms including plants. In 1944 ethylene dibromide, a soil fumigant, was produced. This led to investigation of other bromine compounds and methyl bromide, which is very volatile, came into use for greenhouse soil sterilization. The above are all simple compounds.

A dithiocarbamate, metham-sodium—a water-soluble solid used as a nematicide in seed beds, nursery crops and ornamentals—was one of the first complex chemicals to be introduced. A number of organophosphorus compounds have been shown to have high nematicidal activity. One of the most widely used is aldicarb. Several of these organophosphorus compounds can act on leaf-dwelling nematodes, owing to their systemic action.

MOTHPROOFING CHEMICALS

Clothes moths and carpet beetles cause millions of pounds of damage every year to clothing, rugs, carpets, and upholstery, and in recent years compounds have been developed to render such articles mothproof. Among them are a number of common insecticides such as dieldrin and DDT at low concentrations, which, if sprayed on to the cloth, will keep it moth-free for months or even years. Sodium fluosilicate is another chemical which can be fixed in wool and is an effective agent. Two other recently developed chemicals, 'Eulan' and 'Mitin FF', are stated to provide protection from pests during the lifetime of the fabric. These are both sulphonic acid derivatives. Both are applied to fabrics in a water-soluble form. They combine with wool proteins and are thus 'fixed' in the fibres.

WOOD PRESERVATIVES

Many wooden articles such as construction timbers, telegraph poles, railway sleepers and so on are exposed to attack by a great variety of soil pests including bacteria, fungi of all kinds, and many insects including termites. Zinc and copper salts are often used as preservatives impregnated into the wood, but the most widely-used preservatives are the pentachlorophenols. Creosote, which is a complex mixture of crude coal tar distillates, is a cheap and widely used preservative for timber.

MIXTURES OF INSECTICIDES

The formulation of mixtures of two or more insecticides is now common practice—it is thought that more than 25 per cent of products marketed are of this kind. There are a number of reasons for this. In some cases it is accidental in the sense that the product contains more than one chemical with insecticidal powers and it would be uneconomic and pointless to try to separate them: pyrethrum contains a mixture of pyrethrins, and one of the organochlorine insecticides, toxaphene, is a mixture of several compounds. In other cases the mixing is deliberate, to effect better control—a mixture of pyrethrum and DDT gives a very quick and effective kill. It is of interest here to note that sometimes a mixture of two chemicals will have a more toxic effect than one would expect than if the effect were purely additive. The term 'synergism' has been used for this effect.

Synergism is a widespread phenomenon not only in the pesticide field. It is probable that there are a number of mechanisms involved some of which are highly specific for the reaction occurring: one compound may be practically non-toxic but is active in 'opening up' the organism to attack, e.g. by affecting cell membranes. Another compound may be very toxic but is unable to reach the site of action. Individually, therefore, they would have little effect; together they form a powerful synergistic mixture. In other cases, a compound may inhibit the enzyme which causes the breakdown of the insecticide, and thus the combination of the two is much more active than either of the two individual chemicals would lead one to expect. It may also be noted here that the opposite effect of synergism may occur in mixtures. In other words, the mixture may be *less* toxic than one would expect on additive grounds. Such a phenomenon is known as antagonism and is also fairly common in instances where mixtures of insecticides are used to control a mixed population of insects and thus provide a wider spectrum of activity.

Chapter 5

HERBICIDES

The earliest forms of chemical weed control were concerned with the killing of weeds on roadsides, verges, paths and so on, and the chemicals used were generally cheap byproducts of the newly-evolving chemical industry. Such chemicals included arsenical compounds which were smelter wastes, iron sulphate—a byproduct of the steel industry—salt, and various waste oils. Until the turn of this century practically the only use for chemicals in weed control was as soil sterilants which killed all vegetation with which they came into contact. In other words the chemicals were non-selective.

The best known and most effective non-selective killer, sodium chlorate, came into general use early in the twentieth century and it is still in use today. It kills plants swiftly and efficiently but it has a number of drawbacks. One is that it is very soluble in water, and so moves rapidly in the soil, running down paths on to lawns or 'creeping' into vegetable patches. It also is a highly inflammable substance and in the dry condition it may render organic materials with which it comes into contact—such as plant remains, wood or clothing—highly inflammable too. As a result of this potential danger, commercial products of sodium chlorate may contain a fire depressant such as sodium borate (which is a pretty efficient weed-killer in its own right) or calcium chloride.

SELECTIVE HERBICIDES

A 'selective weedkiller' is one that has a greater effect on weeds than on crop plants; the first was probably copper sulphate. Almost simultaneously, and apparently quite independently, three workers, Bonnet in France, Schultz in Germany, and Bolley in America found in 1896 that a solution of copper salts applied to mixed stands of broad-leaved weeds in cereals would kill the weeds and harm the cereals little if at all. This discovery may well have stemmed from the use of copper salts for the control of plant diseases (see p. 82).

Soon afterwards came the introduction of sulphuric acid for the same purpose. Its selective action depended on the fact that it ran off the leaves of the cereal plants but clung to and burned the relatively rough surfaces of the weeds. In other words it was a 'contact' weedkiller. In the 1930s, thousands of acres of cereals in Britain were sprayed with sulphuric acid, but weed control was still in its infancy. Special equipment was required for spraying and a limited range of weeds was controlled, but sulphuric acid was difficult and dangerous to use and the first real breakthrough came in the late 1930s with the exploitation of the dinitro compounds such as DNOC and dinoseb.

Both of these are contact herbicides, but they are much more selective than sulphuric acid and they readily kill many annual weeds in cereals. During the 1939–45 period, and afterwards, DNOC played a major role in increasing food production. Both DNOC and dinoseb are, however, highly toxic to animals and humans by ingestion, inhalation or absorbtion through the skin, and protective clothing is worn during spraying. As we shall see, DNOC has now been largely replaced—though dinoseb is still used for a number of purposes.

Hormone-like herbicides

MCPA The development of herbicides in their modern form derives from the discovery of the so-called 'hormone-like' herbicides in the early 1940s. These are organic chemicals produced in the laboratory which have certain properties similar to the plant hormone indole-3-acetic acid (IAA), which occurs naturally in plants and which controls many of their vital processes. It has been known for some time that plants produce this hormone, which plays a major part in regulating their growth. But it was not until the early days of the war, when Britain was trying to step up food production, that this knowledge was applied to the killing of weeds.

Three scientists, Slade, Templeman and Sexton, working at the Jealotts Hill Research Station of Imperial Chemical Industries, were investigating the role of hormones in plant growth. They sprayed mixed stands of oats and charlock (a well-known weed) with solutions of naphthylacetic acid (NAA), and found that it killed off the weeds but left the cereal unharmed. Slade and his co-workers realized the importance of their discovery. They tried out NAA against a great variety of weeds, finding that many of them were killed; and against a great variety of cereals, finding that all of them were unharmed. They then turned their attention to chemical relatives of NAA in an attempt to find more potent killers. The result was the discovery that one chemical known as MCPA (a phenoxyacetic acid with chlorine and methyl substituents) was one of the most active compounds and could kill many weeds at very low concentrations.

2,4-D Over the same period three scientists at Rothamsted Experimental Station—Nutman, Thornton and Quastel—by one of those amazing coincidences, had reached almost the same conclusions by a different route.

Certain bacteria termed rhizobia have the power of entering the roots of certain plants (peas, beans, clovers) and causing swellings called nodules. The swellings are not harmful to these plants; on the contrary they are beneficial, for within them the rhizobia can 'fix' atmospheric nitrogen and make a proportion of it available for use by the plant for the formation of proteins.

Nutman and his co-workers considered that the hormone IAA mentioned above was involved in the formation of nodules, and they investigated the effect of applying small quantities of IAA to the roots of certain plants. They found that some plants were killed off by the IAA whilst others, e.g. cereals, remained unharmed.

Like their colleagues at Jealotts Hill they realized that they had a potential weed killer. Looking at related chemicals they hit upon 2,4-D, which was first mentioned by a Dr. Pokorny in 1941 but without reference to a use for it. 2,4-D and MCPA have very similar chemical structures, both being substituted phenoxyacetic acids.

In 1942, the results of this preliminary work by these British teams were communicated to the Agricultural Research Council, who asked Professor G.E. Blackman of Oxford University to initiate a programme of field trials on the findings of the research workers. Both chemicals were potent weed killers and furthermore they were selective, i.e. they did not harm cereals. Even today 2,4-D and MCPA are still the most widely-used herbicides. The two original papers together with the results of the field trials all appeared in one copy of *Nature* (No. 3939, April 28th, 1945) because their publication had been held up for security reasons.

In 1942, two Americans—Zimmerman and Hitchcock—described the use of 2,4-D as a plant growth regulator, but not as a herbicide. Again in the United States, Marth and Mitchell reported in 1944 that it killed dandelions and plantains in a lawn. In the same year Hamner and Tukey, also in the USA, described successful field trials using 2,4-D as a herbicide. It is therefore difficult to establish precedence for the precise discovery of 2,4-D as a herbicide.

Synthesis and activity of MCPA and 2,4-D

Both MCPA and 2,4-D were quickly adopted. From the beginning 2,4-D was almost exclusively preferred to MCPA in the USA. The reverse was true in the United Kingdom. This was dictated largely by the fact that the 'starting' chemical for MCPA, namely orthocresol, is relatively abundant in Britain as a product of coal-tar distillation. The American coal fields were less rich in this product, but on the other hand the production of synthetic phenol, the starting chemical for 2,4-D, was much more plentiful. Both 2,4-D and MCPA are used in almost equal quantities in Europe.

The value of these compounds, apart from the fact that they are very efficient selective weedkillers, is (1) they are cheap to produce; (2) little skill is required in their application because they have a wide margin of safety for all crop plants provided the instructions are followed; (3) they are virtually non-poisonous to man and his animals; and (4) they were produced at a time when farm labour was in short supply and was to continue to be in short supply.

2,4-D and MCPA are absorbed by both root and shoot, and they move rapidly through the plants. This means that the plant does not have to be completely covered by the spray: a few drops can do the trick. The effects are striking. Tumour-like growths, distortion of the tissues with grotesque twisting and contortion of stems and leaves, and masses of minute roots may appear within a week or so of spraying. Internally there is a massive increase in cell numbers and intrusion of certain tissues into others—in other words a chaotic development of growth.

Despite the enormous amount of work that has been done on the subject there are still gaps in our knowledge of how these chemicals work at the biochemical level.

Other related herbicides
Following on the discovery of 2,4-D and MCPA there came the development of a related chemical 2,4,5-T, which has the advantage over the others that it is particularly active against woody plants. It has been widely adopted as a brushwood killer.

In the early 1950s, Professor R. Wain at Wye College, University of London, began looking at some interesting chemical relatives of the phenoxyacetic acids. These were certain phenoxybutyric acids. In theory they should have no effect on plant growth, and yet when they were sprayed on to certain plants such as nettles they caused contorting and twisting in the same way that the substituted phenoxyacetic acids did. On other plants such as celery they had no effect at all. The solution that Wain came up with was interesting. He proposed that nettles and other affected plants had a certain enzyme complex, (a β-oxidase) which converted the phenoxybutyric compound to the corresponding phenoxyacetic compound; thus MCPB was converted to MCPA, or 2,4-DB to 2,4-D. Such plants killed themselves by the products of the conversion. Other plants, including celery, did not have such an enzyme system, the phenoxybutyric compound was not altered, and therefore these plants were unharmed. MCPB was not converted to MCPA; 2,4-DB was not converted to 2,4-D; and since the phenoxybutyric acids are harmless at the doses normally used, the plants remained unaffected. Wain's hypothesis was subsequently borne out by experimentation. This finding opened up a whole new concept of weedkilling based on the enzyme system of the plant. MCPB is particularly useful for killing weeds among certain clovers (because these clovers lack the enzyme system for conversion), and of course for killing nettles in a field of celery. In practice MCPB is often formulated with MCPA.

Although 2,4-D, MCPA and their relatives are very efficient herbicides, and are still the most widely used selective herbicides in the world, they are not effective against all weeds. On some they have little or no effect and as susceptible weeds such as poppies, charlock, runch, shepherd's purse—the common weed species of the pre-war years—are now rarely seen in any well-managed cereal crop, so resistant weeds such as couchgrass, wild oat, blackgrass, chickweed, cleavers and mayweed, came to the fore. This has meant that new chemicals have had to be developed to deal with them. In addition there was, and is, an increasing demand for herbicides to be used on crops other than cereals. As a consequence a wide variety of herbicides is available to the farmer, and all are welcome.

The control of grass weeds
Grass weeds have always been particularly difficult to get rid of, so the introduction by Du Pont and the Dow Chemical Company of TCA in 1947 and dalapon by Dow in 1953 was particularly welcome. Until then it had been relatively easy to kill broad-leaved weeds in grasses but very difficult to kill grasses in other forms of foliage. Both TCA and dalapon are toxic to grass species generally and are usually used in the form of sodium salts at rates per acre much higher than that required of, say, MCPA to kill cereal weeds. Both can be skin and eye irritants (TCA more so than dalapon), and care has to be taken in their use; but they are of low toxicity to man and his stock. Both are useful for killing off perennial grass weeds such as couchgrass—a particularly troublesome weed in Britain and the USA—Bermuda grass and Johnson grass.

TCA is absorbed mainly by the roots, and mixing the chemical with the soil by means of cultivation plays an important part in obtaining successful results. Dalapon may be absorbed through both foliage and roots, and it moves readily through the plant, causing chronic toxicity and killing slowly. Dalapon (like TCA) may act by coagulating plant proteins and thus interfering with their functioning. As such, both chemicals may be directly toxic to all protoplasm but in particular to the enzymes. In addition it appears that dalapon may also interfere with the plant's formation of pantothenic acid (which is one of the B vitamins essential

F   69 to growth and development), since it has been shown that artificially feeding the acid to treated plants may overcome dalapon's toxic effects.

Another herbicide active against grass weeds, though not exclusively so, is aminotriazole, which was introduced by Amchem Ltd in 1954. It moves rapidly in the plant, and applied at the season of declining top growth it can move down into underground stems and so kill difficult weeds like couchgrass and horsetails. Affected plants have a bleached appearance as aminotriazole interferes with the formation of chlorophyll; growth soon ceases (partly as a result of starvation) and the plant dies. The chemical has a low toxicity for animals and man. One of the most troublesome grass weeds in cereal crops is blackgrass (so called because the ripening ears and stems turn dark in colour). It can seriously deplete crop yields, and losses of 50 per cent are not unusual. Its seeds can remain viable in the soil for ten years or more and germinate to give rise to vigorous plants that compete successfully with the cereal. The same chemicals are used for the control of blackgrass as for wild oats.

The control of wild oats is a particularly difficult problem, and they are widespread in grain-growing temperate regions of the world. In their vegetative form they are almost indistinguishable from cultivated oats. Wild oats scatter at harvest time so that with crop harvesting the seeds are incorporated with the cereal grain and are extremely difficult to separate out. A number of chemicals have been developed to deal with wild oats; the most promising so far is a carbamate compound, barban, introduced by the Spencer Chemical Company in 1958. It is sprayed on when crop and weed are at the seedling stage, and it effectively stops the growth of the weed by interfering with the development of its growing points. Two carbamate compounds have also been developed specifically for wild oat control by the Monsanto Chemical Company: di-allate and tri-allate. These compounds act on the oat shoot as it pushes through the soil. Both chemicals are volatile and have to be well incorporated into the soil. Tri-allate is preferred in cereals because of its greater safety in the crop, while di-allate is used for similar purposes in crops such a sugar beet. Good results in cereals have also been reported by using chlortoluron (introduced by CIBA-Geigy) pre-emergence, and by post-emergence spraying with benzoylprop-ethyl (introduced in 1969 by Shell Research Ltd).

Control of resistant broad-leaved weeds
As well as the grasses, a number of broad-leaved weeds have been shown to be resistant to the phenoxyacetic acid herbicides, and as a consequence a number of chemicals have been introduced to deal with them. Closely related are the phenoxy-propionic acids. Mecoprop was introduced in 1957 by Boots Pure Drug Company in Britain as a complementary herbicide to MCPA and 2,4-D, because of its ability to control in particular cleavers and chickweed—two important weeds in cereals that could previously only be killed by contact herbicides. Mecoprop can also deal with most of the weeds controlled by MCPA, and is often used mixed with other herbicides.

Dichloroprop, also a Boots product, is particularly useful because it will control various species in addition to those weeds controlled by MCPA. Fenoprop was developed by the Dow Chemical Company in the USA and although first used against woody plants it has also been used successfully against submerged and emergent aquatic weeds.

Various benzoic and phenylacetic acids such as 2,3,6-TBA, introduced by the Du Pont de Nemours Company, and dicamba by the Velsicol Chemical Corporation, are used against weeds resistant to MCPA and 2,4-D. Ioxynil and bromoxynil were developed by May & Baker Ltd and by Amchem Ltd for the control of chickweed, mayweeds and some polygonums in cereal crops. They are in the main contact herbicides of moderate mammalian toxicity.

Control of weed seedlings
Related to ioxynil and bromoxynil is dichlobenil, the herbicidal properties of which were discovered both by Shell in Britain and by Philips in Holland. It is very potent when applied to the soil for the control of germinating weed seeds and it is widely used for the control of weed seedlings in rice crops. Dichlobenil is volatile but is moderately persistent in the upper layers of the soil, and it is of low mammalian toxicity. An interesting compound has been developed by Shell in Britain: chlorthiamid, which is converted to dichlobenil in the soil.

Two substituted urea compounds—monolinuron and linuron have been especially developed by Hoechst in Germany (linuron also by Du Pont in the USA) for weed control in potato crops. These herbicides are absorbed through the foliage as well as

71

through the roots, and they control many seedling weeds as well as those germinating for some time after application. Another, chloroxuron, is a CIBA-Geigy Ltd product used in a similar way and for a similar purpose in strawberry beds.

Three closely related uracil compounds, bromacil, lenacil and terbacil from Du Pont in the USA, are all primarily absorbed from the soil by the roots but they can also enter through the shoots. They are toxic to a wide range of grasses and broad-leaved plants, interfering with the photosynthetic process, and are very persistent in the soil. Bromacil is used for the control of weeds in citrus orchards, lenacil for the control of weeds in sugar beet and terbacil for selective use in sugar-cane and a variety of sub-tropical fruit crops. They are also used at higher concentrations for total weed control.

Two major groups of herbicides which are 'total' weedkillers in the sense that they will kill almost all plants with which they make contact are also important 'selective' weedkillers when they are applied in the proper way to the appropriate crop. These are (1) the substituted ureas, which were first introduced by Du Pont in 1951 and (2) the triazines, introduced by J.R. Geigy in 1956. These two groups are the most widely used herbicides after the phenoxyacetic acid group.

(*1*) *The substituted ureas.* These include in addition to those previously mentioned fenuron, monuron, diuron and neburon. The first three are used in a variety of circumstances, e.g. on railway tracks and industrial sites as non-selective total herbicides. Although they are slow-acting, they are very persistent in the soil, and depending on the amounts used (10–30 lb per acre) they may keep the treated area free of weeds for more than a year. In addition they do not move readily in the soil (they are very insoluble) so that they do not contaminate adjoining land. At low doses (0·25–3 lb per acre) they may be used selectively on certain crops, particularly those with deep roots such as cotton, sugar-cane, pineapple, lucerne, gooseberries and raspberries. Diuron and neburon are particularly used in this way. Because of their low solubility, and also because they are readily adsorbed on to soil colloids, they remain in the top one to two inches of soil. Since this is where most weed seeds germinate, they are rapidly killed off. On the other hand, those crop plants having deep roots are not affected by the chemicals. Fenuron is sometimes used for the control of woody plants and as 'spot' treatment on otherwise very persistent weeds such as docks. The main action of this group is on the photosynthetic system of the plant.

(2) *The triazines*. A large number have been developed in recent years by J.R. Geigy. These include simazine, atrazine and propazine. Like the substituted ureas they have low solubility and they are very persistent in the soil. In massive doses simazine and atrazine are used for non-selective weed control, either alone or mixed with other herbicides. Like the substituted ureas, and in similar circumstances, they are also used as selective weedkillers, particularly in tree and bush fruits, woody ornamentals, forest nurseries, sugar-cane, pineapple, brambles and so on. Chlortorulon is used against blackgrass. The killing action is likewise probably through an effect on the photosynthetic system. Propazine is used as a selective herbicide in millet. Atrazine is widely used in maize crops because maize is resistant to this chemical. The mechanism of resistance is of great interest; it is brought about by an enzyme system in maize which rapidly breaks down the chemical. The substituted ureas and the triazines have a very low mammalian toxicity.

Paraquat

A most interesting group of herbicides was introduced in 1958 by Plant Protection Ltd, UK. These are the quaternary ammonium compounds—paraquat and diquat, which act similarly on all plants including grass, killing the aerial parts by contact action within a day or two of application. They are also translocated to a limited extent within the plant. They are widely used for the destruction of existing growth before reseeding, thus offering a chemical alternative to ploughing. Indeed paraquat has been nicknamed the 'chemical plough' and it is of particular value on hill land which is inaccessible to the plough. Another interesting feature is that these compounds are very rapidly immobilized in the soil, being adsorbed on to soil colloids before being broken down. This means that it is possible to sow seed within a day or two of chemical treatment. Diquat is used for killing off potato foliage before potatoes are harvested. The effect, as on other plants, is a drying out or dessication of the foliage. The compounds are dependent upon light for the full expression of their desiccant and contact effect. Paraquat and diquat are also effective in killing submerged aquatic weeds because of their quick contact effect. These compounds should be used with great care, avoiding skin contact and inhalation as much as possible.

A new class of herbicides represented by glyphosphate was introduced by Monsanto Chemicals Limited in 1971. They are used as foliage sprays and their speed of action is enhanced by increased light intensity. Glyphosphate controls a wide variety of annual and perennial weeds including couchgrass, docks, dandelions, thistles and brambles. It is inactivated immediately on reaching the soil. It obviously has a bright future in clearing weeds from orchards, vineyards, forestry and industrial situations.

Growth regulators

The discovery of the plant hormone indole acetic acid by Frits Went in the Netherlands more than 40 years ago has led to the establishment of a whole new group of plant growth substances— the auxins. Since that time a number of new hormones have been found including the gibberellins (a few drops of which may cause dwarf plants to make spectacular growth) cytokinins (which are involved in cell division), absciscic acid (an inhibitor of plant growth) and ethylene (responsible among other activities for the ripening of fruits). It seems clear that the growth of a plant is in large part made up of an interplay between these (and perhaps other) hormones. Many benefits have accrued to agriculture and horticulture from these fundamental findings. The auxins have led to the hormone weedkillers, rooting powders, and chemicals preventing premature fruit drop; the gibberellins are used to speed up the malting process, and to increase the size of grapes; ethylene compounds (e.g. 'Ethrel' by Amchem Limited) are used for the ripening of fruits.

Other purely synthetic growth substances are also used in agriculture such as daminozide (introduced by Uniroyal Inc in 1962) a plant growth retardant used for the control of vegetative growth of fruit trees and the modification of stem length and shape of ornamental plants; maleic hydrazide (US Rubber Company) which retards the growth of grass, hedges and trees and prevents sprouting of potatoes in store; and chlorflurecol-methyl (introduced by E. Merck in 1965) a general growth retardant. There are many others: some are available for the shortening of straw length of cereals without reducing the yield. The moulding of plant growth by the use of chemicals will be one of the major developments in agricultural science in the years ahead.

Weed control in special situations

These are grouped according to the Weed Control Handbook.

Cereal Crops The use of herbicides for the control of weeds in cereals, as we have seen elsewhere, has led to a marked change in the weeds present. Poppies are now so scarce as to merit exclamation when they are seen, and in many areas—though not as many as there should be—yellow charlock has all but gone. This change has been brought about by the use of MCPA and 2,4-D. But as we have noted, the virtual disappearance of susceptible weeds has led to the spread of resistant ones so that new weed problems have to be faced. MCPA and 2,4-D are still however our most widely used herbicides—weeds do not succumb too easily to attack, and still have hidden reserves of seeds in the soil; and there are still plenty of foci of infestation from non-sprayed fields.

Among the resistant species, some annual broad-leaved weeds such as speedwell and corn marigold, and annual grasses such as blackgrass and wild oats, are very troublesome. So too are perennials such as creeping thistle, docks, field bindweed, couchgrass and bent grasses. Despite the development of a wide range of chemicals such as those mentioned earlier in this chapter, many problems still exist which tax the ingenuity of farmer, adviser and chemist alike.

Row Crops Among crops grown in this way are french beans and broad beans, sugar beet and fodder beet, brussels sprouts, cabbage and cauliflower, carrots, kale, maize, onions, peas, potatoes, swedes and turnips. Developments in chemical control in the last ten years have revolutionized the growing of these vegetables.

One technique that is commonly applied for weed control in these crops is termed 'pre-emergence'. Just before the crop comes through the ground, and while it is still protected by the soil, the seedling weeds which have arisen on the prepared seed bed are sprayed with a contact herbicide. The two essentials for success are first that the majority of weed seedlings should emerge before the crop (and cultivation of the soil for a few days before planting usually ensures this), and second that the chemical used should be an efficient contact herbicide that does not persist in a toxic form in the soil. Among the chemicals used for this purpose are paraquat, diquat, triflurin (introduced by Eli Lilly and Co., 1960), propachlor (by Monsanta in 1965) and dinatramine (by Borax Consolidated in 1971).

Some chemicals such as prometryne (J.B. Geigy, 1962), aziprotryne (CIBA-Geigy, 1967) and Methazole (Velsicol Chem. Corp., 1968) can be used both pre-emergent and post-emergent on certain crops, and desmetryne was introduced by J.R. Geigy in 1964 as a post-emergence translocated spray for brassica crops.

In order to assist the lifting of potatoes it is common practice to spray the foliage with a desiccant some 10 days before harvest. Dinoseb, diquat, sulphuric acid and others are used for this purpose.

Perennial crops other than grass The most important perennial crops on which herbicides are widely used are apples and pears, bush and cane fruits, strawberries and bulbs. A less important group comprises rhubarb, asparagus, and herbaceous ornamentals grown for flower and foliage. The use of herbicides for weed control in these crops is a recent development and is mainly due to the development of simazine as a soil-applied herbicide.

As mentioned earlier, simazine can keep the soil clear of weeds for a considerable time but will not harm the perennial crops which are so much deeper rooting. The introduction of paraquat and diquat has also revolutionized weed control in this area and the glyphosphates will also make a big impact. In this case they can with care be sprayed on to the weeds, and because of their rapid inactivation in the soil they do no harm to the crop plants. Dalapon and TCA are also used for the control of grass weeds.

Grassland The term grassland covers a great variety of forage crops, including (1) annual, temporary and permanent crops of grasses and legumes (such as clovers, lucerne and sainfoin) sown by man, and (2) those areas that make up the 'permanent grass' and 'rough grazings' which cover about two thirds of the area of Britain. The weed problems encountered vary not only from crop to crop but also from area to area, e.g. in the early stages of sown grass we encounter annual weeds such as chickweed and groundsel which are really more characteristic of arable crops, whereas in permanent grassland perennial weeds such as buttercups, docks and thistles are common; and the weeds of permanent grassland on the chalky soils of the Downs are very different from those of permanent grassland on the acid soils of the Scottish uplands.

Much can be done in the way of weed control on grassland by good management such as controlled grazing, draining and liming. Nevertheless chemicals must also be used, particularly for the control of perennials including those mentioned above.

76

For some, such as buttercups and daisies, MCPA and 2,4-D are very effective. Others such as docks and bracken are highly resistant and the introduction of asulam by May & Baker Ltd looks like being a very welcome breakthrough. Both paraquat and diquat can be used where there is a need to kill off all vegetation before sowing, as can dalapon and aminotriazole.

The choice of chemical depends upon the botanical composition of the sward. Dalapon and aminotriazole are more effective than paraquat on perennial grasses. But both of these chemicals persist in the soil and an interval of six weeks is required before resowing, whereas paraquat and diquat are quickly inactivated.

Turf for sport and amenity

The control of weeds, including moss and undesirable grasses, is an essential part of turf management. Indeed, until the advent of selective weedkillers, good management, drainage, aeration, feeding, and cultivation were the beginning and end of the production of good turf. They are still of prime importance, but the help of weedkillers has been greatly appreciated. Herbicides such as diquat, paraquat, aminotriazole and dalapon are widely used for seed bed preparation. 2,4-D and MCPA will deal with most of the common weeds in the developing or mature sward, whilst mecoprop and dichlorprop are being increasingly used for specific problems such as the control of white clover.

Regular treatment over several years with MH will change the ground composition and the finer-leaved dwarf grasses (such as red fescue and smooth-stalked meadowgrass, are increasing at the expense of coarser grasses.

Forests and forest nurseries

The forest nursery is rather like a market garden. The crop is of high value, but it can be checked in growth by a heavy weed infestation and so every effort is made to keep the beds weed-free in the early stages. As a consequence high-cost herbicides may be used, provided they are highly efficient. The tree seedlings are grown in the beds for one or two years before being transplanted. The soils most suited to the growth of tree seedlings are generally lighter and more acid than agricultural and horticultural soils, and this has an important bearing on the weeds that grow there. Thus two important weeds are annual meadowgrass and sheep's sorrel. Since conifer seed is sown on the soil surface and covered with coarse sand, it has little if any protection from applied herbicides; and as a consequence we find that there are not many satisfactory pre-emergence herbicides. Probably the best is vaporizing oil which volatilizes within a few hours of treatment; paraquat is also widely used for this purpose. White spirits may be used when the seedlings are six weeks old. After transplanting in rows, simazine can be used very successfully to keep the rows free of weeds.

When the young trees are finally put out on to their permanent sites the main aim in weed control is to prevent weeds smothering them before they have a chance to grow. On peaty hill land there is seldom a weed problem, as single furrow ploughing is practised. The natural vegetation buried under a plough-ridge is slow to recover and the young tree has grown away before it does. Nevertheless, dichlobenil has proved to be a considerable help, particularly in suppressing the competitive effects of bracken.

However, on fertile hill land and lowland hill sites weed control is necessary because of the speed at which the natural vegetation regenerates. Chlorthiamid, used in granular form, gives an acceptable control of grasses and herbaceous weeds; paraquat, dalapon and atrazine may also be used. It is sometimes necessary to clear such sites of trees or shrubs such as hazel, oak and rhododendron before planting and this is generally done by using mixtures of 2,4-D or 2,4,5-T and ammonium sulphamate together with mechanical methods. In general, there is not however a high use of herbicides in forestry.

Weed control in land not used for crops

Total Control Herbicides are used for the total control of weeds on railway tracks, petrol installations, fire-breaks, paths, playgrounds, and around farm buildings. In order to rapidly kill existing vegetation aminotriazole, dalapon, MCPA, 2,4-D and 2,4,5-T, paraquat or diquat and glyphosphate may be used during the growing season. In most places it is desirable to prevent regrowth for considerable periods, so fairly long-lasting persistent herbicides are used such as triazine, simazine, diuron, monuron, sodium chlorate and borates. They are mainly root-absorbed, their action is slow, and they are best applied in late winter or early spring. In some cases mixtures of rapid- and slow-acting types may be applied during the spring and summer months when the vegetation is growing. Maintenance treatments may be applied from time to time to prevent reinfestation. Combinations of herbicides have proved to be very effective. A mixture of diuron and bromacil prevents regrowth of grasses for over a year; similar results have been obtained with diuron plus aminotriazole; and even two years after treatment with bromacil plus terbacil, there may be very little regrowth. Metribuzin plus bromacil has also been reported to be very effective.

Partial Control There are a number of situations where a partial control of vegetation is desired—for example on highway verges and motorway and waterway embarkments, where the aim is to reduce the height of the vegetation. This is no easy matter; and the methods most commonly used are mechanical ones. But costs are high, and the use of herbicides is tempting. The danger in using them is that even selective herbicides may remove some species altogether and agreement has been reached between the Ministry of Transport and the Nature Conservancy in Britain on the use of phenoxyacetic acid herbicides for the control of roadside vegetation.

Aquatic weed control

For centuries water weeds have been controlled by hand-cutting methods to keep rivers and brooks free-running, and to prevent ponds from becoming overgrown. With the reduction in manpower for this kind of work machines have been developed, but in some situations they are not entirely satisfactory and there is an increasing use of herbicides. But herbicides must be used with care because the very mobility of water makes it an excellent transporter of these chemicals, and the risk of adverse side-effects at the treatment site and downstream must constantly be borne in mind. Of particular importance ate the effect on human safety and water supplies; the danger to farm animals and to fish and other important aquatic animals; the possible hazard to adjacent crops; the risk to wildlife nesting-sites; and the danger of damage to industrial plants. There are several Acts of Parliament covering the introduction of poisonous substances into streams.

Among the herbicides used are dalapon for the control of rushes, reeds and sedges, and 2,4-D for the control of many broad-leaved weeds. Both of the above are sprayed on the foliage. Submerged and floating weeds such as Canadian water weed, water milfoil, crowsfoot and pondweed are treated by introducing the herbicide into the water to form a dilute solution. Among the

79

herbicides used in this way are diquat, dichlobenie and chlorthiamid. Maleic hydrazide is used on banks to encourage the growth of a short grass sward. All of the above have been cleared under the Pesticides Safety Precautions Scheme (1973)—see p. 174.

Chapter 6

FUNGICIDES

The use of chemicals for the treatment of plant diseases goes back to classical times, when we find Democritus in 470 B.C. suggesting that the press cake left behind after making olive oil should be sprinkled on plants affected by the powdery mildew. Dr. Horsfall, a well-known American plant pathologist, made the interesting observation that it took man some 2,400 years to rediscover (in 1931) that vegetable oils are very effective in the treatment of mildew-affected plants.

Inorganic sulphur compounds
Sulphur is one of the oldest of the fungicides. As far back as 1803 recipes were available for its use and by 1846 it was being extensively used in English glasshouses for the control of powdery mildew of the vine. It is still an important fungicide and may be used in various forms. Finely ground it is used as a dust; it is inclined to 'clump' but this can be overcome by the addition of some inert material such as kaolin. Such finely-divided sulphur can also be used as a spray provided a 'wetter' such as soap is added. A third method of use is by volatilization, i.e. the conversion of the sulphur directly into a gas by means of heat. This method was in use in the 1850s when hot water pipes of greenhouses were painted with sulphur. More sophisticated methods have since been developed, including the blowing of hot air over molten sulphur. In 1852 a M. Grison, a French gardener, boiled lime and sulphur together and so 'Eau Grison' or lime sulphur was born. It is now produced commercially and is a very effective fungicide particularly against fruit tree diseases. It may cause

81

some damage to sulphur-sensitive plants such as tomatoes and grapes, particularly in hot dry weather.

Copper

Though copper sulphate had been used as a fungicide much earlier, for example against 'bunt' disease of wheat (in which the grains are replaced by black fungal spores) it was not until Millardet accidentally observed its efficiency against the vine downy mildew in 1882 that its possibilities were recognized. Millardet, a professor in the University of Bordeaux noticed that grape-vines along the roadside in the St. Julien district of the Medoc retained their leaves at the end of October when all the others had been defoliated by mildew. On enquiry he was told that it was customary to sprinkle the vines near the roadside with copper sulphate and lime to discourage pilfering. It occurred to Millardet that as well as checking pilfering it was also checking the disease and he obtained confirmation of this when he carried out tests near Margaux in the following year. He published his results in 1885 and by 1887 great success was reported in the control of the disease. The mixture of copper sulphate and lime became known as 'Bouillie bordelais' or 'Bordeaux mixture' and by 1888 it was being used against potato blight—as it is used today, with minor modifications. It is of interest to note that it had been noticed even earlier, in 1846, that blight was not present on potatoes near a copper smelter in Wales while those further away were seriously affected. But it is one thing to notice something; it takes genius to apply it.

In common with most other fungicides, Bordeaux mixture does not cure the disease. Its function is to prevent it. Sprayed on to leaves it will poison fungal spores as they germinate. In the case of potato blight it is now possible to predict an outbreak of the disease by studying the weather conditions in advance, temperature and humidity being particularly important. Farmers are warned some days in advance when conditions are right for blight to occur in a few days' time, and the potato foliage is sprayed with Bordeaux mixture or one of its more modern successors.

Phenolic compounds

A number of phenolic compounds which we have already noted are active as herbicides and insecticides are also fungicidal. The best known is DNOC (dinitroorthocresol), but because of its high toxicity for plants it is used for spraying on orchard soil to kill overwintering fungi. Two dinitro compounds have been developed which have a low toxicity to plants: dinocap and binapacryl, both of which are used as protectant sprays against the powdery mildews which attack the leaves of various plants.

Dithiocarbamates

A major development took place in 1931 when Du Pont introduced a new fungicide, which was known as thiram. This was a dithiocarbamate, and it led to the development of a completely new series of fungicides. The discovery of the fungicidal activity of the dithiocarbamates illustrates once more how important the 'hit or miss' method can be. Scientists investigating the mode of action of sulphur as a fungicide drew a theoretical parallel between the action of sulphur in vulcanizing rubber and its fungicidal effect. Their theory was not accepted; but it was very productive in that it led to the examination of many sulphur compounds, used in the rubber industry to accelerate and control the vulcanization process, as potential fungicides. Thiram was one of these rubber accelerators which became useful principally as a seed dressing.

Other dithiocarbamates followed in quick succession. They are now the most widely-used and most versatile of all the organic fungicides, ranking with 2,4-D in the herbicide field and DDT in the insecticide field. They are used for spraying on foliage, for seed treatment, for killing fungi on turf, as textile and paint protectants and even for the treatment of fungal skin diseases of animals and man. Zineb and maneb are used on a variety of plants as foliar sprays. Metham-sodium is a soil-applied chemical which controls not only soil fungi but also eelworms and weeds. The dithiocarbamates have low mammalian toxicity—a factor which makes them very acceptable general fungicides.

Heterocyclic compounds

Some of the best fungicides contain heterocyclic rings. One of these is captan, which was first introduced in 1949 by the Standard Oil Development Company and is used as a foliage protectant, as a seed protectant and as a postharvest dip for fruits and vegetables. Another fungicide from the same company is folpet, which is used in rather similar situations but is also particularly effective against many of the powdery mildew diseases. Captafol was put on the market by the Chevron Chemical Company in 1961; it is used against the blights of potato. Since 1961 a number of other heterocyclic fungicides have also become available.

Quinones

In the unending search for substances that will kill fungi one obvious place to look is inside plants themselves, to determine if there are any chemicals or groups of chemicals active against fungi. Plant physiologists have shown that one such group is the quinones. Some quinones occur naturally but others can be synthesized in the laboratory. Two such quinones have been developed in this way; and they are used commercially as fungicides. One of them, chloranil, is used mainly for the treatment of seeds and bulbs (particularly of flower and vegetable plants) but also as a spray and a dust. Dichlone is used for similar purposes but also as a protectant spray for certain blights, fruit rots and cankers. Both are products of the US Rubber Company.

Organotin compounds

Although inorganic tin compounds show no activity against fungi the same is not true of organotin compounds. Indeed these are among the most powerful killers of fungi that we have. Many of them kill not only fungi but also many other living things and cannot be used selectively. The development of these pesticides came about as a result of collaborative work between the International Tin Research Council in London and the Institute for Organic Chemistry in Utrecht.

So far the biggest use of an organotin compound is for the protection of certain transparent plastics from breakdown in the presence of light; but it was found that this compound, a trialkyl tin, could also protect materials against attack by organisms. It is used in the rot-proofing of fabrics and as a constituent of anti-fouling paints which are used to protect ships' bottoms from being overgrown by seaweeds and barnacles. In recent years, the Hoechst Chemical Company have found that the triphenyl tin compounds, whilst retaining their ability to kill fungi, are not nearly so poisonous to mammals; and they have introduced fentin acetate which, like fentin hydroxide (introduced by Philips-Duphar), is used to control leaf spot of sugar beet and celery, blight of potatoes, and a range of fungal diseases of coffee, banana and sugarcane. Higher yield-increases have been found to result from their use than have followed control with copper compounds and some have suggested that tin has a stimulating effect on the growth of plants. Others however consider that the increase is due to its having a less toxic effect on the plants than copper.

Antibiotics
Antibiotics are chemicals produced by one micro-organism which are active against other micro-organisms. Antibiotics have, of course, found their main use in the medical field but some have also been used for the control of diseases of plants. Among the most important of them is streptomycin. Tetracyclines, cycloheximide, and griseofulvin have on occasions been used to a much lesser degree.

Streptomycin is produced by an actinomycete—*Streptomyces griseus*—and is used as a spray against a number of bacteria which cause plant diseases. It is also used as a protective dip for seeds and potato seed pieces before planting. Streptomycin is also active against a few important fungi. Cycloheximide is produced by the same actinomycete, and is indeed a by-product of streptomycin production. It is used in the control of a number of fungal diseases such as the powdery mildews. The tetracyclines are also produced by members of the *Streptomyces* group, and some of them such as oxytetracycline, chlortetracycline, and tetracycline are used against some of the bacterial diseases of plants. Griseofulvin was first isolated in 1939 from growths of *Penicillium griseofulvum* and in the medical field it was later shown by Dr. J.C. Gentles in Glasgow University to be active against fungal diseases of humans. It causes fungal hyphae to twist and curl, and it is also said to interfere with chitin formation in mosquito larvae. It has, however, been little used in the practical control of plant diseases or mosquito control.

Seed treatment of fungal diseases
A number of important plant diseases are carried in or on the seed. A number of these, known as the 'smuts' and 'bunts', are particularly troublesome. The spores of the smut fungus lodge in the grain covering and germinate there; the fungus which develops grows into the grain and makes its way up the developing shoot. It does little or no harm until it reaches the 'ears' of the

G

plant where it produces black spores, smuts, in profusion—which completely replace the grains. The whole of the head may have a dark, sooty appearance and is not only economically worthless but also serves as a source of infection for other plants. Seed treatment of cereals with fungicides has now become commonly-accepted practice against this and other diseases.

Seed treatment is one of the oldest methods of disease control, its origins going back to the seventeenth century when wheat seeds were soaked in salt water before planting. The process was known as 'brining' and it is thought to have developed as the result of an accident at sea. It is said that about the year 1670 a sailing vessel loaded with wheat encountered a storm and ran aground near Bristol. Farmers living along the coast collected some of the grain washed up on the shore; and, finding that since it was saturated with salt they could not use it for flour, they planted it. Perhaps someone noted that the plants grown from the seeds from the sea remained free of smut disease whereas normally there were a good many smutted plants in the crop. One man with a particularly fertile mind might have put this to the test the following year, and finding it to be true the practice would soon be generally adopted. This, of course, was long before it was known that smut or indeed any other plant disease was caused by a living organism.

In 1748 Jared Eliot described the method:

If a farmer in England should sow his Wheat dry without steeping in some proper Liquor, he would be accounted a bad Husbandman. Mr. Ellis directs to put in a quart of Salt into as much Water as is sufficient to cover two Bushels and a half of Wheat, to which add two Quarts of slaked Lime, but in the Wheat after you have well stirred the Brine, let the Wheat steep twelve Hours, then draw off the Water, spread the Seed after it hath dreined half an Hour, sift upon it dry Lime stirring it about, which will make it dry enough for immediate sowing. I have not room to name the Advantages they say they have from this Practice. When I did it this Year I used Tyde water; this saves Salt. Some direct to make the Brine very strong, so as to bear an Egg.

The use of an egg as a hydrometer for testing salt solutions appears in the report of the Commissioner of Patents in 1844 in a summary of prevailing practices as regards seed treatment for the control of 'the smut which is sometimes found in wheat called dust brand or pepper brand. The substances used are sulphate of copper or blue vitriol, wine, common salt, wood ashes, lime water and sometimes arsenic'.

Copper sulphate was first used in 1761, but application difficulties were soon encountered. The seed had to be soaked in copper sulphate solutions—no easy job for the farmer, particularly as the seed had to be spread out to dry before planting. The 'heap' method was easier: the grain was piled up and sprinkled with a solution of copper sulphate, shovelled to ensure mixing, and finally spread out to dry. The method was widely used, but was unsatisfactory because of injury to the grain. More successful was the dry treatment with copper carbonate fungicides, begun in about 1917.

Formaldehyde was first employed as a disinfectant in 1888. It was introduced as a seed treatment in 1895 and was widely used in the USA. In Europe it was not much used until the First World War when copper and sulphur became scarce. For some time thereafter it was one of the most popular seed disinfectants until the development of the organomercury compounds, of which a wide range is now available for use on vegetable seeds as well as on cereal seed grain.

The development of mercury compounds for the control of fungal diseases stems from the use of mercuric chloride for killing bacteria. Owing to the fact that mercury is very poisonous to animals and humans, most mercury compounds are used for seed treatment; only a few are used as foliage sprays and then primarily for diseases of turf. Mercuric chloride was first tried out as a dressing for cereal grain in 1890, and in a more detailed fashion in 1915 when it was observed that it did prevent disease of the grains. It is however a very toxic substance, and was quickly replaced when in 1915 I.G. Farbenindustrie in Germany introduced a liquid organic mercury compound—chlorophenol mercury—which was much less poisonous, and was used for the control of bunt. 'Germison' was introduced in 1920 by the Saccharin Fabrik A.G., and its success led to the introduction of hydroxymercury chlorophenol by the Du Pont Company in the USA in 1924. Organomercury dusts for seed treatment were introduced in 1924, but they posed hazards in use. Nevertheless ethyl mercury chloride was produced by IGF and a compound

87

containing a mixture of phenyl mercury acetate and ethyl mercury chloride by Imperial Chemical Industries in Britain. 'Slurry' treatments consequently became popular, and methyl mercury dicyandiamide from Sweden came onto the market in 1930. Since then there has been a steady stream of organomercurials. They are generally used as seed dressings, either as dusts or as slurries for a great variety of crops including cereals, cotton, rice, flax and many other crops, for bulbs of all kinds, for fungal diseases of turf, and for incorporating with papers and paints to give them fungus-resisting properties.

Systemic fungicides

It is not surprising that it has taken much longer to develop systemic fungicides than it took to develop systemic insecticides. It will be recalled that the objective is to find a chemical which will not only be taken up by the plant without harm, but which will nevertheless poison any fungus feeding on it. Since fungi are also plants (albeit not green plants) the two physiological systems have more in common than have a plant and an insect. It is more difficult to find a chemical that will differentiate between a green plant and a fungus than to find one that will differentiate between an insect and a plant. Nevertheless it has been done, and the development of systemic fungicides promises to revolutionize plant protective measures. The first to be introduced was probably triamiphos in 1960 (by Philips-Duphar). There are now many different classes of organic chemicals being used as systemic fungicides. Among them are the oxathiins, of which carboxin (introduced by Uniroyal Inc in 1966 for the control of loose smut of barley) is the best example; the pyrimidines, of which there are two compounds—dimethirimol and ethirimol—active against the powdery mildew fungi; the benzimidazoles which show a wide spectrum of activity, both benomyl and thiabendazole being toxic to a wide range of fungi including the powdery mildews, the apple scab fungus, and the grey mould fungus which attacks all sorts of plants and fruits; the thiophanates whose two fungicides, thiophanate and thiophanate methyl, are used extensively on various fruit and vegetable crops particularly in Japan and in Italy; and finally the morpholine fungicides, of which tridemorph is probably the best known, which are specifically active against the powdery mildews.

Systemic control of rust, smuts, and powdery mildews, as well as the fungus *Botrytis cinerea* which is a common cause of grey mould on soft fruit, is as we have noted now well advanced; but there is still a great need for a systemic that will control the Phycomycete group of fungi which causes some of the world's most devastating plant diseases such as blight of potato.

Systemic fungicides can be applied to assist in the control of certain of the deep-seated smut diseases afflicting wheat and barley. It appears that the fungicide does not go in through the seed coat but enters the young roots as they push their way out of the seed. A benomyl soil drench is recommended for the control of *Verticillium* wilt of cotton, while cucumber mildew is controlled by soil treatment with a variety of products. The comparatively recent introduction of systemic fungicides in agriculture has brought about dramatic improvements in the control of many hitherto very intractable plant diseases. Spray treatment of leaves by systemics is now common practice; but other techniques are required for the control of the deep-seated fungi that cause Dutch Elm disease and silver leaf disease of stone fruit, both of which are situated deep within the trunk. The former disease is currently being successfully treated by boring holes into the trunk and injecting a systemic fungicide. Such treatment has been used in the past for the introduction of trace nutrients in which trees growing in certain situations may be deficient. It seems clear that as new and improved systemic fungicides are developed in the future, the whole picture of plant disease control may be revolutionized. Their development and impact may well equal the effect that the introduction of 2,4-D and MCPA have had on weedkilling and that DDT had on insect control. One worker in the field (Dr. M.C. Shephard) has said, 'I would hazard a guess that the balance between systemic and non-systemic fungicides is likely to parallel the use of antibiotics and poultices in medicine.'

Resistance

We have already noted (Chapter 4) that some insects have become resistant to some of the insecticides. A similar problem is now cropping up with regard to fungi and certain systemic fungicides. In the short time that has elapsed since the introduction of the systemic fungicides there have been more reports of resistance to fungicides than ever before in the whole history of fungicide usage, e.g. cucumber mildew is already resistant to both benomyl and dimethirimol. The reason for this is not clear but it may be that fungicides of the older type such as copper or sulphur killed by interfering by a great many vital processes in the fungus which had therefore little or no chance of developing resistance to this multipronged attack. The systemic fungicides on the other hand cannot hit at many systems because some of these are common to the green plant too. In other words a chemical with such a multi-pronged attack would also kill the green plant instead of moving harmlessly within it. It is therefore possible for systemics to hit at only one vital system within the fungus—one that is different from the green plant. The fungus therefore has a greater opportunity of developing resistant strains. Whatever the reason, it is a worrying development and one that pesticide scientists will swiftly attempt to overcome.

Chapter 7

THE APPLICATION
OF PESTICIDES

The formulation of pesticides
Most pesticides are applied as liquid sprays, but some are used in
the solid form as granules or dusts. Though there are a few exceptions, they are mostly applied in such low dosages that they
have to be mixed with a solid or liquid carrier to ensure even
distribution. The commonest (and cheapest) liquid carrier is
water and most pesticides are formulated specially in order that
they can disperse readily in it. Some which do not dissolve in
water are instead prepared so that they form emulsions—i.e.
they are marketed in oil together with a suitable emulsifier, so
that when they are added to water they form an oil-in-water type
suspension. Wettable powders consist of a finely-divided solid
pesticide, a suitable filler such as clay, and dispersing agents
whose purpose is to give a stable suspension of the solid particles
when mixed with water.

One very important aspect to be considered when deciding the
formulation of any pesticide is the addition of surface active
agents known as 'stickers' or 'wetters', which enable the pesticide
to spread evenly over the plant or insect surface, thereby over-
coming the water-repellent nature of that surface. With some
pesticides the addition of these substances not only assists in
spreading, but also makes for a more efficient penetration by the
pesticide. To increase the efficiency of some pesticides, a mineral
oil or some other organic solvent may be recommended in place
of water.

In the manufacture of dusts the pesticide is either mixed
directly with a powdered carrier or is dissolved in a solvent and

sprayed on to the carrier. Granular formulations are becoming increasingly important for soil application. They consist of granules or pellets of various forms. They too are usually prepared by spray impregnation: a solution of the pesticide is sprayed on to a pre-formed granular carrier; or by agglomeration, starting from a powdered mixture of chemical and carrier.

Soil-applied pesticides

Some pesticides are not sprayed onto crops or weeds but are applied directly into the soil. In such cases the character of the soil can have a marked effect on efficiency: a clay soil will hold on to the chemical more readily than will a porous sandy soil. On the other hand, even distribution is more easily effected in a sandy soil. It has been found that granular formulations are particularly well suited to soil application. They are convenient to handle and do not give rise so serious drift hazard; they tend to rebound from foliage, so most of the herbicide reaches the soil; and it is possible to modify granules in such a way as to control the rate at which the herbicide will be released into the soil.

Spraying equipment

New pesticides and methods of applying herbicides are continually being developed. Spraying machinery is an active subject of investigation in the pesticide field, and recent years have seen major advances. One of the commonest forms of equipment for small-scale projects is the portable knapsack sprayer. It can carry up to three gallons, and the spray may be applied by gravity or with the aid of a pump. The sprayer is slung on the body of the operator who directs the lance or boom delivering the spray on to foliage or soil. Some sprayers incorporate a small petrol engine and a fan to provide an air blast. They are widely used, particularly for the application of pesticides to trees and bushes.

The majority of machines in use in agriculture are mounted on tractors, and have a capacity up to 100 gallons. The pump is usually mounted on the power take-off of the tractor and connected to the spraying unit by suction and delivery hoses. Spraying contractors and farmers with large acreages to spray find it economic to use larger spraying machines. These are usually similar in principle to the tractor-mounted machine, but have a greater tank capacity which is achieved either by the use of a trailed unit or by mounting twin tanks on either side of the tractor. A few years ago 21-foot booms were normal; now the boom is more likely to be 30 foot, with some contractors' machines going to nearly 60 foot. This creates an acute driving problem: the driver either overlaps or underlaps. An adequate form of marking is the answer—failure to mark may lead to wasteful use of chemical or to strips of the crop being unsprayed.

There are a few specialized machines for use on railways and roadsides. These have high-output nozzles, arranged so that the swath-width can be varied as work proceeds by turning them on and off individually. Compound nozzles are sometimes used instead of a boom. Heavier equipment suitable for travelling in soil can also be used, and the speed of travel when spraying is generally much faster than when using crop sprayers.

Application of granules

Granular formulations of herbicides have so far been used mainly on industrial sites, but also to a limited extent agriculturally for narrow band treatment or 'spot' treatment of small patches of particularly difficult weeds. Insecticides for the control of mosquito larvae in overgrown pools and streams are frequently applied as granules, which are more effective than sprays for penetrating the canopy. Similarly, granules are less likely than spray in agricultural application to attach themselves to the crop, and they are therefore particularly suitable where soil application is required in a standing crop. Hartley & West (see further reading), however, point out an exception to this practice which is used to advantage. In the maize plant, the leaf clasps the stem, forming a funnel in which granules collect. It is in just this region that the stem borer makes its entrance; and the collection of insecticide granules in the funnel is much more effective than a spray, because the funnel is not watertight. Another situation where granules are very effective is in the control of root-attacking larvae, where the granules can be implanted close to the root.

A limitation to machine design is that it is difficult to obtain even distribution of granules of mixed size and density so that the manufacture of granules of uniform size is important. But there are a number of machines available. The smallest and simplest consists of a hand-held container fitted with a device which allows a controlled flow of granules to escape down a tube. Application in bands is useful in crops grown in rows. A metering device consisting of a fluted roller is also in use. Granules falling into the flutes are carried round and dropped into the distribution tubes; the dose is regulated by changing the length of roller into which the granules can feed. This equipment can be used both in hand-propelled and tractor-mounted forms. For large areas centrifugal distributors similar to those used for fertilizer distribution (with with better distribution control) are being developed. This equipment can be used both on the ground and from the air.

Aerosols

In most aerosols the active component, in this case pesticide, is mixed with a carrier liquid of low boiling point. The most commonly-used carriers are the fluorochlorocarbons, which have practically no smell, low toxicity for mammals, and are not inflammable. A very fine spray can be produced by discharging the liquid plus pesticide through a very fine orifice. Since the liquid is compressed below its boiling point in the container, it issues as extremely fine droplets, since it boils when it meets the air. The method is expensive, the cost of the chemical being only a fraction of the cost of the container; and only small containers are used as they must withstand high pressure in storage. They are used in houses, small greenhouses, etc., because of their great convenience. The small size of the droplets means greater penetrating power, and insecticide aerosols are very efficient killers.

A modification is the foam dispenser which is widely used domestically e.g. for shaving cream. Foaming is achieved by using a lower pressure and a larger orifice, and foam-stabilizing agents are incorporated. The method is used with herbicides for 'spot use' on a small scale where, on lawns, a little foam containing the weedkiller can be put on to a particularly troublesome daisy or dandelion—thus obviating the need to spray the whole lawn. Butane is commonly used as a carrier.

Smoke generators

Insecticidal and fungicidal smokes are generally used in confined spaces such as greenhouses, but may also be used in forests. The smokes are produced from special generators based on the machines used during the Second World War for the production of different coloured smokes for identification purposes. It was found during the War that complex organic dyes would stand up to the smoke process; it has subsequently been found that complex insecticides will also resist breakdown when treated in this way. Thus we can obtain 'smoke' of azobenzene for mite control; pyrethrum, gamma-BHC, parathion and DDT for insect control; and captan for fungus control. The advantages of the smoke method over spraying is that the smoke can penetrate into small crevices and otherwise inaccessible areas.

In the generator, the insecticide or fungicide which is to be vaporized is mixed (a) with an oxidant such as sodium chlorate and (b) with combustible material—such as sugar—capable of generating a large amount of hot gas (water vapour and carbon dioxide). The pesticide is protected from destructive oxidation by the fact that slightly less sodium chlorate is used than is necessary for complete oxidation of the sugar (which is more reactive with sodium chlorate than is the pesticide). Another important factor is that the chlorate and sugar are thoroughly mixed before adding the insecticide. Dilution of the vapour with air to produce a fine smoke is brought about automatically by the turbulence set up by the high velocity of the jet of hot gas. The preparation of such a smoke generator is, of course, no job for the amateur since sugar and sodium chlorate can form a highly explosive mixture.

Fumigants

Related to the use of smokes for pest control is the use of fumigants. The method goes back a long way to the burning of aromatic herbs for the supposed protection against disease, and is now accepted as an important method of pest control in certain situations. It involves the use of a toxic chemical in volatile form and is employed in enclosed spaces such as greenhouses, or under 'tents' or tarpaulins. Tents are used in the treatment of citrus trees—the tree being covered by an air-tight tent into which is introduced the required amount of fumigant. Polythylene and PVC are now widely used as cover for soil fumigation, replacing the tarpaulin of times past.

Fumigants are used in empty greenhouses or warehouses for disinfestation purposes before re-use; for the treatment of grain or other storage products; for soil treatment; and against burrowing animals such as rabbits and moles (see Chapter 8).

The first case is a relatively simple one: the fumigant has easy access to all parts of the building. There is no question of injury to plants or other produce within it, and a period of ventilation after treatment gets rid of traces of the fumigant. The treatment of grain is more difficult in that the movement of the fumigant is restricted by the grain, and the fumigant must be of such a nature that there is no tainting of the grain or any harmful residue left within or on it. Soil fumigation is particularly difficult, since the movement of the fumigant is even more restricted than in grain, and the constituents of the soil may absorb and immobilize the chemical.

The earliest efficient fumigant to be used on a commercial scale was probably hydrocyanic acid, used in 1886 against the scale insects on citrus trees in California, and this is still much used today. It is produced by the action of a strong acid on sodium cyanide, and packages of sodium cyanide are marketed in zinc-foil containers ready for dropping into jars of acid. Liquid hydrocyanic acid may also be obtained, and granules of calcium cyanide which yield hydrocyanic acid gas when moistened with water have become increasingly employed. Although the gas can injure plants it has been found that it is less harmful to them in the dark. Consequently cyanide fumigation is now carried out at night. It is the most effective fumigant for empty buildings, but has poor penetrating powers and is little used in the treatment of grain. It is widely used, however, by quarantine authorities in the treatment of plant materials—which must be in the dormant condition or injury may ensue. Cyanide is a very acute poison, and great care must be exercised in using it. Although to some it has an odour of burnt almonds, others cannot smell it at all and this increases the dangers in its use. In practice an irritant such as chloropicrin is often added to the gas so that anyone exposed to it will have due warning.

Naphthalene, derived from coal tar, is a white crystalline substance with a characteristic smell. It has a scorching effect on plants and is not widely used for greenhouse fumigation. But it has long been known to possess insecticidal powers, and was widely used as 'camphor balls' or 'moth balls' for the protection of clothing.

Nicotine is used as a fumigant in glasshouses against aphids. It may be heated over a small fire or spirit lamp, or alternatively strips of paper are soaked in saltpetre, dried and then impregnated with a mixture of linseed oil and nicotine. The strips are hung up and nicotine vapour is evolved.

Methyl bromide, a colourless liquid boiling at 4·5°C, is used extensively for the 'sterilization' of nursery stock. This sterilization is a necessary requirement in many countries before such stock can be imported. It is also an important soil fumigant, being particularly active against fungi. It is much less toxic than hydrogen cyanide, but its poisoning effect is much more insidious and therefore less likely to be noticed particularly since it has little or no smell. As with hydrogen cyanide, chloropicrin is often added as a warning irritant.

Soil fumigation
The aim behind soil fumigation is the destruction of such pests as wireworms, eelworms, fungi and weed seeds in the soil.

Carbon disulphide was first used in 1872 against phylloxera of the vine, by injecting it into the soil. It is cheap, highly volatile and it kills insects. It has been combined with a number of other compounds, such as carbon tetrachloride, to reduce its fire hazard in transit and to slow down its loss in soil by evaporation. Its use has not spread far beyond the vine-growing areas of France.

Ethylene dibromide, DD and chlorodibromopropane are all soil fumigants. The latter compound is the least toxic to plants and it is used in established citrus groves. The other two are only used in soils before planting. All are active against eelworms and insects. They are generally applied in a localized manner, injected into field soil in a continuous line, or as a series of spot applications.

Another soil fumigant is chloropicrin, which has already been mentioned as an irritant added to HCN and methyl bromide. It is more mobile in moist soil than hydrogen cyanide but it is also more stable; its main loss is not by evaporation but by breakdown in the soil. It is also used for fumigating dry seeds in store, though it is very lethal to germinating seeds. There must be adequate ventilation after use to avoid taint.

Coal tar compounds—such as cresylic acid, dichlorocresylic acid and dichlorobenzene—are used to some extent as soil

97

sterilants. Various halogenated hydrocarbons are produced commercially, particularly for use against eelworms. Formaldehyde is an effective soil fungicide but has little or no effect on insects; and it is difficult to clear the soil after use.

Aerial spraying

One of the advantages of aerial spraying is the saving in labour costs. From the air a plane can spray about 500 acres in a day. Another very real advantage is that there is no physical damage to the crop by tractors. And thirdly, operations can be timed over a shorter period so that the spraying can be done when weather conditions are suitable. Aerial spraying has now developed into a highly commercial operation; and much fundamental scientific work has gone into the devising of apparatus and the operating of the machines for maximum efficiency.

Originally aerial spraying services began as 'solo' ventures. The end of the Second World War gave an impetus to the development of aerial spraying. When many young pilots found themselves demobilized with a flying licence and little else to support themselves, some of them acquired small aeroplanes and offered their services to the farmer. The operations were often crude but at least it was a beginning. The 1950s saw the development of a much more professional approach. Companies were formed offering their services as contractors, equipment was specially designed for fitting to the aircraft, and today there are many companies operating throughout the world. Some of them spray more than 50,000 acres in a year.

Investment in this type of operation is high. Aircraft cost around £10,000 each and there are considerable overheads. It has been estimated by one operator that it costs £27·50 per hour to operate a full team with aircraft. The charge to the customer is around £1 per acre and the cost of the chemical has to be added on to this. Spraying fields from the air is really a team job with men on the ground as well as in the air. The field to be sprayed is fitted out with fluorescent marker strips which are placed at intervals across the flight paths. These act as guides to the pilot so that there is a minimum of overlap and no misses. Spraying is carried out at heights related to wind speed. The chemical is located in tanks and a pump is used to force the spray out through booms fitted beneath the aircraft wings.

In crop protection, pesticides have long been applied using high volumes of water to dilute the pesticide and thus ensure adequate coverage of the crop. Usual amounts have ranged between 100 and 500 gallons per acre. The drawback of having to transport all this liquid around is obvious, and the development of more efficient spraying equipment has meant that less dilution is necessary—a great many pesticides are now applied at the rate of 7–25 gallons to the acre.

Although this is perhaps an acceptable level on the ground it is still too high for use in aircraft and even in the early days of aerial spraying the standard practice was to fly as low as possible over the crop applying 2–5 gallons per acre in swaths of 40–60 feet. One of the limitations on the use of aircraft is the volume of spray that can be carried. It is therefore of great interest that much thought has been given to the development of what has been termed 'ultra-low-volume', or 'waterless spraying'. This has been developed for a number of reasons. Water is not a particularly good carrier for pesticides because (1) the small spray droplets evaporate rapidly in their passage from the sprayer to the target, causing them to become lighter and to drift; (2) most target surfaces such as the cuticle of plants, the coverings of insects and the cell walls of fungi are water-repellent; and (3) some pesticides will not dissolve readily in water.

There are benefits to be derived by doing away with water. The pesticide need not be formulated in such a way as to become water-dispersible, the droplets of spray retain their size and thus reach the target site—the pesticide in an oil formulation for example penetrates more readily—and, most important of all, the operator has merely to transport the concentrated pesticide and not the enormous quantity of water for dilution purposes.

The first modern synthetic insecticide applied in an undiluted form was used in 1957 to control an agricultural pest on the plains of Somalia; the pest was the desert locust and the chemical was DNOC. As we have noticed elsewhere, the desert locust may occur in massive swarms which may contain about 30 million locusts per square kilometre and may migrate 2,000 kilometres from their breeding grounds. R.D. MacCuaig and others of the Anti-Locust Research Centre in London, in co-operation with the Chemical Defence Experimental Establishment at Porton Down near Salisbury, developed a technique by which a 20 per cent oil concentrate of DNOC was sprayed from an aircraft on to a settled swarm. At a little over 570 gallons per square mile the mortality was nearly 100 per cent. Dr. MacCuaig describes how by a close investigation of droplet size, angle of spray and so forth it became possible to effect an almost total kill with as little as 20–40 gallons of diazinon per square mile; thus a 30-square-mile swarm could be destroyed by only two or three sorties by each of three spray planes. Previously a total of 50–60 sorties had been needed, and the task was often impossible because the swarm moved out of range in the time needed. Because of the problems of toxicity with concentrated diazinon, a switch was made to 20 per cent dieldrin concentrate, which however also proved to be toxic—on one occasion sheep and goats were killed and ultra-low-volume spraying consequently suffered a setback, indicating that great care had to be exercised.

Tests have also been carried out against a number of pests with other pesticides in concentrated form such as DDT and malathion. In the early 1960s the Tropical Pesticides Research Institute and the Desert Locust Control Organization for East Africa took part in a combined operation to improve crop spraying in Africa. In Tanzania DDT oil concentrates applied as fine aerial sprays at 0·5 gallons to the acre from Micronair atomizers were as effective against bollworm on seed beans as were emulsion concentrates applied as coarse sprays at 2 gallons per acre from conventional boom and nozzle spray gear. In the USA the low volume technique was first used in 1962, when it was found that one pint per acre of dieldrin applied by Micronair atomizers was as effective as one gallon per acre in controlling rangeland grasshoppers.

Helicopters are also used to some extent for crop-spraying purposes, and they have a definite advantage over fixed-wing aircraft in mixed cropping situations. This is because, at very low speeds and heights, the down-draught from their rotor blades tends to improve the covering by pesticide of the lower sides of the leaves of the particular crop. This is all the more desirable since a number of pests are normally to be found on the undersides of the leaves.

Herbicide drift
Valuable crops can be destroyed by drift from herbicidal sprays and it is important that all possible steps should be taken to eli-

minate this risk. The growth-regulator types of herbicide such as MCPA and 2,4-D are particularly dangerous in this respect because they are so active at very low concentrations; and some crop plants like turnips, swedes, tomatoes and lettuce are particularly susceptible. The manufacturers are constantly working on formulations that will reduce drift to a minimum, but much can also be done by spraying only when conditions are suitable e.g. when there is no movement of air towards a susceptible crop or on warm still days when convection currents rise from the crop.

Combined pesticides and fertilizers

It is obviously of great advantage to the farmer both in time and money if he can combine two or three operations into one exercise. Much thought has therefore gone into the possibility of mixing two or more herbicides together to increase the efficiency of kill; of combining herbicides with insecticides so that both weeds and insects can be killed at one spraying; of combining herbicides, insecticides and fungicides in one spray so that plant diseases can be controlled in addition; and of the combination of any one, two or all of them with fertilizers.

Success depends on a great many factors. Chemicals that are mixed together must be chemically compatible—they must not mutually inactivate each other, and since the total chemical content will be high they must be readily soluble. Their effects on the plants sprayed must be complementary, not antagonistic—the efficiency of one chemical should not be markedly lowered by the presence of the other. Ideally one chemical should increase the efficiency of the other: the spraying must take place at a time which is near to optimum for the desired effect. This is probably the most difficult factor to satisfy—it may be desirable to kill weeds when they (and presumably the crop) are at a seedling stage of development; but this may not be the ideal time to spray for disease control. Some insects, too, may be best hit when the crop is mature—and this may not be the best time for fertilizer application. One major problem is the difficulty of applying the correct rate for both materials of a standard mixture. Despite the difficulties, however, some progress has been made.

The combining of different herbicides is common practice: for example, proprietary formulations containing propham plus

H

chlorpropham plus fenuron have been employed for several years for weed control in red beet; and MCPA has been combined with a great many chemicals, as with MCPB for weed control in clover crops, with ioxynil for spraying cereals, and with 2,3,6-TBA and atrazine for total weed control. There are many more examples.

A number of combined herbicide and insecticide sprays are available. Herbicides have also been successfully combined with nutrients, e.g. manganese sulphate with dinoseb for the combined control of marsh spot (manganese deficiency) and weeds in peas; and MCPA or 2,4-D with a nitrogenous fertilizer such as urea, nitrate of soda, or sulphate of ammonia for nitrogen deficiency in various plants.

Chapter 8

VERTEBRATE PESTS
AND THEIR CONTROL

The Rabbit

Among the most common of the vertebrate pests is the rabbit. Originally a native of south-west Europe and north-west Africa, it was probably introduced into Britain by the Normans in the twelfth century and was kept in carefully guarded warrens as an important source of food. Towards the end of the nineteenth century, however, rabbits became numerous—so numerous that they became pests. The reason for this rapid upsurge in numbers is not known.

During the nineteenth century they were introduced into Australia, where, lacking natural enemies, they bred rapidly and spread throughout the country. They competed successfully with the sheep for grass because despite their relatively small size five rabbits can eat as much as one fully-grown sheep. In most of the countries to which they have been introduced—Britain, Australia, New Zealand, South America, the United States, etc.—they have become serious pests, voraciously devouring crops. In Britain, for example, it was shown in the early 1950s that rabbit devastation was costing the country something like £50m. a year. The toll in Australia during the same period was undoubtedly of a much higher order.

Until the coming of myxomatosis to Britain, a number of methods were employed to try to keep the rabbit in check. Snares of all kinds and gin traps were abundant until the latter method was outlawed. Rabbit shooting was a common sport, serving the double purpose of pest control and providing meat for the pot. Rabbit warrens were dug or ploughed up; rabbits were harried by ferrets which were deliberately introduced into their burrows—they were either killed by the ferrets or were netted or shot as they sped out. Poison gases were pumped down their burrows. And yet they went on multiplying rapidly until 1953.

When myxomatosis struck in 1953 (see Chapter 9) it was thought by many that the rabbit threat was over in Britain and elsewhere: during 1954–55 over 90 per cent of all rabbits in Britain were eliminated. Since then, there has been some reduction of virulence of the disease. It would appear that the relationship between the virus, the flea and the rabbit is changing but interpretation is at present a very difficult matter. The disease is still present and intervenes periodically when rabbit numbers tend to increase. The partial recovery of the rabbit population has meant that recourse must again be made to trapping, shooting and chemical methods. The most effective method is to introduce poison into the burrows, and since there is little likelihood of affecting other animals there is no need for a selective killer. Hydrogen cyanide is the gas that is commonly used for this purpose; and the most widely used form is a powder which liberates the gas on contact with moist soil. It is usually blown into burrows with a dust gun. In order to ensure safety in storage it is usually mixed with magnesium carbonate and anhydrous magnesium sulphate. In Britain such a product is marketed by Imperial Chemical Industries under the trade name 'Cymag'.

The Rat

Whereas the rabbit is regarded by the general public with some affection, although it does an enormous amount of damage, the same is not true of the rat. There are two important species—the black or ship rat which is largely confined to ports and large towns, and the brown or common rat which is active in towns and in the countryside.

The brown rat is a burrowing animal living in earth banks, haystacks and holes in the ground in the country, while in cities it nests in sewers, drains and the basements of buildings. Although it is less associated with disease than the black rat (which harbours the flea which carries the dreaded plague) it far outstrips it as a plunderer of food in Britain (where the black rat is now rare), and does an enormous amount of damage particularly to stored foodstuffs such as grain. The life of the brown rat is closely connected with that of man—it follows him all over the world from Alaska to the Antarctic. It probably came to Britain from the East on trading ships, around 1730. It is a prolific breeder—a female begins breeding when she is only about 80 days old and then produces up to 50 young in a year although in practice more than 99 per cent die before they become adults and most adults do not live more than a year. This rapid rate of reproduction and turnover makes rats difficult to control. They feed on almost anything, eating not only vegetable materials of all kinds but also all kinds of animals—mice, poultry, ducks and so on. They migrate *en masse* when their homes become overcrowded. In many places including Britain the black rat has been ousted by the brown rat as a result of competition for food and living space.

Trapping, chasing with dogs or shooting has little effect on numbers of rats so it is hardly surprising that they have become a target for pesticides. In times past two plant products, red squill and strychnine, were extensively used. The former is obtained from the bulb of the sea leek, a member of the tulip family, and a number of powerful toxins combine to make this a very potent poison. Interestingly enough, its effects seem to be almost confined to mice and rats; but this is not because other animals are immune to the poison. It is because it is a strong emetic and is regurgitated by all animals except mice and rats before it can do any harm. The poison affects the heart and produces convulsions and respiratory failure; it decomposes fairly quickly in the body of the victim.

Strychnine is obtained from the seeds of the plant *Strychnos nux vomica*. It is most effective against rats and mice, although it is generally toxic to warm-blooded animals. It produces convulsions by interfering with processes in the central nervous system.

Both fluoroacetamide and sodium fluoroacetate are useful against rodents inhabiting large tracts of land; and although these chemicals are highly toxic to other mammals, they are particularly suitable when used by trained operators in special environments such as sewer systems. They cause convulsions in the rat. Barium carbonate has been widely used as a rat poison in the United States. It has no smell, and by its action it causes the rat to seek water and thus die in the open. Another compound which is slow in producing death and likewise causes the rat to leave its normal haunts and to seek water is thallium sulphate. Death is due to respiratory failure, but it has a high toxicity for man and has to be used with the greatest care.

However, the demand for most other rat poisons dropped almost to zero with the introduction of a compound known as warfarin (discovered by the Wisconsin Alumni Research Foundation in the USA), which seemed to make all other rat poisons obsolete. It had long been known that spoiled sweet clover is poisonous to cattle, and in 1941 Dr. K.P. Link and colleagues at Wisconsin discovered that the active ingredient in the sweet clover was dicoumarin which interferes with the action of vitamin K in the body and reduces the coagulating properties of the blood so that trivial injury can lead to fatal haemorrhage. It was shown that a daily dose of 2 mgm was fatal to rats, whereas dogs survived daily doses of 50 mgm. Humans can tolerate much higher levels. Link and his group examined the anticoagulant properties of a range of coumarin compounds and selected the forty-second on their list, WARF 42/3, as the most promising as a rat poison. This compound was to become known as warfarin.

It is one of a number of similar chemicals which are now used to control rats, mice and other rodents. They act by interfering with the clotting mechanism of the blood and they probably also cause a breakdown of blood vessels. Rats are very susceptible to low doses, whereas other animals are less affected. Very few rodents survive this chronic form of poisoning and they become too ill to feed before they have taken a lethal dose. This combination of high toxicity for rats and high safety for other animals made it almost the ideal rodenticide.

Then a major snag developed. Rats were becoming resistant to warfarin—and to other similar anticoagulants. The first case of resistance was reported in a rodent pest of sugar-cane in Guyana and then, surprisingly, in a man in California—the latter after examining his blood-clotting factors when he had been on therapeutic doses of warfarin for thrombosis. The first case of resistance in Britain was discovered in 1958 among rats on a farm just outside Glasgow. The area in Scotland now affected by resistance is more than 800 square miles, and there are also large areas in England and in Wales. Warfarin resistance is also present in the mouse. There are now many areas in the world where there are resistant rats. Furthermore, the resistance is hereditary, so that when resistant rats breed the offspring are resistant too. It appears that resistance is due to the effect of a single dominant gene. This means that rats with the gene for resistance will survive warfarin treatment, and if they breed together they will produce mainly resistant offspring. Some of the young rats will wander off and themselves breed, thereby introducing resistance elsewhere. This decrease of effect has meant a renewed search for other rodenticides.

Norbormide looks promising. It is very toxic to rats, which die 'in their tracks' soon after eating a small dose; death appears to be painless. The remarkable thing about it is that its killing action seems to be confined to rats. Even mice are unharmed, and birds have eaten large amounts without effect. Britain's biggest manufacturer of warfarin, Ward Blenkinsop Ltd, has recently discovered a new chemical which is claimed to be effective against warfarin-resistant rats. This new substance, 'sorexa CR', is a mixture of calciferol in corn oil and warfarin in oatmeal, and field trials have proved to be very successful.

The Mink

The American mink is one of the most valuable fur bearers and has been bred commercially in the USA since 1866—on large farms, ranches and in back yards. In 1951 the pelts of two million ranch-reared mink were sold on the United States market. In many places mink have however escaped from their farms and colonised the surrounding countryside, attacking birds and fish. From Scandinavia there have been reports of their turning their attention to salmon, for mink are equally at home on land or water. The first British mink farm was established in 1929 and there are now several hundred farms in the country, some of them carrying 4,000 or more mink. Soon after the establishment of these farms some animals escaped and since 1956 they have been reported breeding in England, Scotland and Wales. They are ruthless killers and one is said to be capable of killing up to fifty hens in a night. Ranch mink have been bred in a variety of colours: brown, beige, blue, cream, white. These may all be found in the wild but the dark-brown animal is the most common. They are generally under 4 lb in weight in the field. The litter number varies from two to six; the young attain adult size in about four months and can breed the following year. They swim and climb well. It is feared that as mink increase in numbers they may become a serious menace, and as yet no very successful control measures have been devised, although shooting and trapping are being widely used in an attempt to contain the numbers. More than 5,000 have been caught in Britain in the past decade.

The Squirrel

The red squirrel is native to Britain but the introduction of the grey squirrel from America in the nineteenth century has led to a decline in numbers, and the grey squirrel has thrived over most parts of the country. Both do serious damage to young trees in forestry plantations, and beech and sycamore are especially vulnerable. For a time there was an attempt to control the grey squirrel in Britain by the government paying for each tail that was produced. This was a costly business but there was no indication of any reduction in the number of squirrels and the scheme was abandoned. Warfarin is now used to attempt to control them, though the dose has to be higher than that now used for rats.

The Mole

The natural home of the European mole seems to have been woodlands, but as these became scarce so it has moved out to the fields. Moles live almost wholly underground, seldom coming to the surface. By their subterranean activities they throw up soil heaps which reduce the value of grassland and encourage the spread of weeds. Indeed the value of an affected area of pasture, as grazing land, can be reduced by as much as 50 per cent and mole-hills can damage hay-making machinery. In addition, by eating worms moles reduce the numbers of this beneficial creature. A mole cannot survive more than a few hours without feeding and when earthworms are plentiful it can store them. It bites off the tip of the worm's head-end, twists the body into a knot and pushes it into a cavity in the soil. These stores may include thousands of worms, but if left too long the worms recover by growing new heads and moving away. But as well as eating earthworms moles may feed on pests such as leather-jackets, wireworms and cutworms. A single mole will eat 40–80 lb of food a year, foraging on average over a tenth of an acre. In times past they were trapped for their skins, but there is now little market for these.

The mole is another burrowing animal which can be treated by means of hydrogen cyanide. In addition there are other methods, e.g. aluminium phosphide tablets are inserted into burrows. They release a very poisonous gas, phosphine. Neither hydrogen cyanide nor phosphine have been shown to be very successful. Another method is to present the mole with poisoned bait; for this purpose strychnine is commonly used. In Britain this is the only pest control use for which strychnine is permitted and it is only issued against a permit granted under the auspices of the relevant government department. The earthworms are dusted with it, and the treated worms are then dropped back into the runs. The mole is a voracious feeder and it takes the bait readily, particularly as it has no sense of smell. Two to three such treatments are usually sufficient to kill all moles in the treated area, and if the bait is carefully laid then no other mammals or birds will suffer.

The Coypu
The coypu, or nutria, is a large rodent introduced into Britain from South America for fur farming in 1929. From the escapes which soon followed the coypu became feral around 1932, and was well-established in Norfolk by 1943. It was not, however, until the late 1950s that numbers increased sufficiently in East Anglia for farmers and others to be concerned about the damage to crops and to river- and dyke-banks. Young sugar beet plants were being killed by coypu biting off the growing tops, and other root crops and cereals were also attacked. The coypu is semi-aquatic, and it makes large holes in river-banks. Where rivers and dykes are above the level of the surrounding land, as in parts of East Anglia, there is the danger of flooding should the banks be so breached.

In England the coypu has few natural enemies other than man and the weather. The hard winter of 1962–63 killed an estimated 80 per cent of the whole population. There was a recovery, however, and the Ministry now meets the cost of coypu control on a pound for pound basis. Systematic trapping has been successful but there are no grounds for complacency.

The Seal
Both the common seal and the grey seal are found round the shores of our northern waters, both being particularly common

round the Scottish coasts. About three quarters of the world's population of grey seals are indeed to be found around Britain. Both live largely on fish: cod, halibut, herring, mackerel and particularly salmon—they are notable pests of salmon nets. Unfortunately they do a disproportionate amount of damage by injuring and leaving more fish then they can eat, and they also rip the nets. Owing to the grey seal being a protected animal its numbers have increased dramatically, so that a number of pups in the Faroe and Orkney Islands have been culled in recent years.

The Pigeon

The wood-pigeon, of which there are about 5 million in Britain, can do an enormous damage to field crops. From being a harmless rarity to the end of the eighteenth century, the wood-pigeon has become one of the commonest and most destructive pests of agricultural land in parts of Europe. It was primarily a bird of the woods, but with the spread of agriculture it changed its habits. Cereal grains are its most important food, followed by peas and beans, and the bird is very adept at digging up and consuming new-sown grain. In winter it attacks greens, brussels sprouts and clover and it has been calculated that wood-pigeons cost British agriculture somewhere in the region of £2m. per year.

Their town relative, escaped domestic pigeons (which are descendants of the rock-dove), foul buildings with their droppings—making expensive cleaning operations necessary. Until fairly recently the Ministry of Agriculture sponsored a scheme whereby it subsidized the price of cartridges for shooting wood-pigeons, but although lots of pigeons were shot it seemed to have little effect on total numbers and the scheme has now been given up. Many sportsmen and cooks regretted its passing, but it is doubtful if it was of any value to the farmer.

Deliberate poisoning using treated bait is illegal because of the danger to other wildlife and to humans who might eat the pigeon corpses. Interesting research has, however, been pursued into the use of stupefying baits treated with narcotics. Pigeons eat this bait, and topple over in a drugged state; they can then be collected up and painlessly destroyed. Licences allowing the use in England and Wales of stupefying baits against town pigeons and house sparrows were first issued to servicing companies and other operators in 1959 and 1963 respectively; and in 1969 farmers were first licensed to employ baiting techniques to catch wood-pigeons. The method has a number of advantages in addition to humane destruction, because any other birds which happen to pick up the narcotised bait can be left to recover and in a short time are as lively as before. Even this side-effect can to some extent be avoided if desired, by treating only large grains such as maize or beans which cannot be eaten by most other grain-eating birds.

The search is progressing for efficient narcotic agents. The one used in most control operations to date is alpha-chloralose, a product formed from chloral hydrate and glucose. This substance was selected following extensive laboratory and field trials. Used against house sparrows, especially those infesting buildings, it has proved to be reasonably effective; but against pigeons it has certain disadvantages, the most important being that it is rather slow-acting, there being a delay of some 50 minutes between the bird taking the bait and the alpha-chloralose having the desired effect. In this period, the birds may have scattered and be out of reach of the collectors. In addition, alpha-chloralose has an initial 'exciting' effect. A number of other drugs have been tested but to date none seem to provide the perfect answer.

The Starling
Starlings also foul buildings with their droppings, roosting as they do in enormous numbers in cities. Interestingly enough, however, they did not move into towns until the 1890s (the first recorded in London was in 1894). They still feed out in the country, and although they eat cereals and may damage fruit they do also eat many agricultural pests such as wireworms and other insects.

It is well known that starlings form communal roosts from autumn to spring and although it may seem otherwise to town dwellers only about 1 per cent of the winter population roosts in urban situations, the majority favouring reed beds, scrub or woods in rural areas. The winter starling population in Britain has been estimated at about 37 million; and some roosts observed may contain about $1\frac{1}{2}$ million birds. In woodlands, starlings prefer to roost in the dense growth of relatively small conifers (in particular spruce), hawthorn, hazel and rhododendron bushes. A Forestry Commission enquiry has shown that the damage that they do is twofold. First, the combined weight of many birds may break off small branches and cause permanent disfigurement of the tree when leading shoots are damaged. Second, the accumulation of droppings—which may exceed one foot in depth on the ground—kills off plants and may in severe cases kill off the trees. In some cases as many as 5 per cent of trees may die when a roost is occupied for one year. In addition, timber merchants sometimes refuse to handle the trees because of the fouling by the starlings.

Because of their noise, filth and smell they are not tolerated near houses and farmsteads; they uproot sprouting grain; and in cherry orchards they do a great deal of damage, eating and despoiling the fruit. Roosts in the vicinity of airfields are also a great hazard when flocks of birds cross the flight-path of an ascending or descending aeroplane: in the United States in 1960, 62 people lost their lives when the aircraft in which they were travelling ran into a flight of starlings and crashed.

Trapping and shooting methods are used, but they do not have much effect. One method which is fairly successful is the use of smoke, which can be produced by burning oily and paraffin-soaked rags. Hand clapping, shouting, football rattles, dustbin-lid banging and fireworks have all been employed from time to time. The use of model hawks suspended from balloons is a method of bird control which is based on the innate fear of birds of prey by small birds, and in Holland some success was claimed by flying a dummy hawk suspended from a motor-driven arm attached to a tall pole.

But this method was given up in favour of what has been called 'bioacoustics'. It appears that when starlings are disturbed they give out a distress call which warns all other starlings within earshot and causes them to fly off. Use has been made of this call in order to disperse starling flocks. The technique was pioneered in the USA in 1954 by two scientists, Frings and Jumber. They captured some starlings, held them by the legs or wings so that they emitted distress calls and made tape recordings of them. They found that when these tape recordings were played through a loudspeaker they had a similar scaring effect upon roosts of starlings. Investigations into the application of this method have been carried out in various countries: in Britain and in Holland in cherry orchards; in Germany and in Switzerland to disperse roosts of starlings in vineyards; in Tunisia to protect olive groves. During the period 1960–8, officers of the British Ministry of

Agriculture, Fisheries and Food used distress call techniques to shift starling roosts from 33 sites; of these, only two failed. After the roosts had been dispersed the sites remained clear for variable periods of time. In two cases the birds returned to the original roosts within two weeks, but in thirteen cases roosts remained unused for the rest of the season.

Success has also been reported using ultra sound—a note so high that it is imperceptible to the human ear, but apparently easily heard, and disliked, by the starling. Using such a method, starlings which were such a nuisance in Glasgow have all but been cleared from the city.

Individual buildings can now be made almost bird-proof by the use of strips of plastic material attached to the ledges where the birds usually roost. These strips do not harm the birds, but the yielding jelly-like material makes them very insecure and they soon depart from a building that has been adequately treated. Buckingham Palace is one of the buildings that has been successfully treated in this way.

The Sparrow

The house sparrow, although it also eats insects, is a pest in many countries: not only because of its habit of nesting in stores and bakehouses, but also because, being avid grain eaters, sparrows can quickly strip a seed crop and beat down the stems. They also consume considerable amounts of grain in store. Shooting and various bird-scaring devices seem to have had little effect and the problem has yet to be solved on the farm. In bakeries and the like, however, sparrows have almost been eliminated by the use of narcotics. A suitable bait is laid early in the morning of a weekend when the building is not in use. A few hours later the sleeping sparrows are collected and humanely destroyed. The drawback of movement away from the site (which we noted with pigeons) does not apply here, so it is unimportant if the narcotic is slow-acting.

The Gull

In some parts of the UK gulls are considered to be pests. One promising method of control is to dip their eggs in a mixture of oil and formalin. The birds continue to incubate the eggs, which do not hatch, and they may not breed for a whole season. If the

113

eggs are removed the gulls lay a second clutch. Work is also in progress on the use of stupefying baits against gulls.

According to one very experienced worker, the most significant finding to emerge from vertebrate pest control (excluding rats and mice in many habitats) has been the conviction that damage should be prevented on a local basis and that attempts to control population over the whole country are usually likely to fail.

Chapter 9

BIOLOGICAL CONTROL

The success of a pest may be measured by its ability to attack other living things such as food-plants and animals, even man himself, more efficiently and speedily than the pest itself can be attacked by other organisms. There is a constant struggle going on in Nature, and man's survival is dependent upon his ability to tip the balance in his direction. Failure to do so would mean that pests would gain the upper hand, with disastrous results.

Among the insects, plants and other living things a similar biological struggle is constantly in progress. All organisms including pests are attacked by parasites that feed upon them. These parasites which prey on pests are often called predators, and the very basis of 'biological control' is to encourage such predators or to introduce new ones to attack the pests. In other words, man has sought out some of these 'natural' enemies and has turned them to his own use. It may be noted that under natural conditions biological control is the normal state of affairs in that it does control most potential pests.

Historical aspects
No-one knows when we first started using predators to control pests but the practice goes back a very long way. Since ancient times Chinese citrus growers have placed nests of the predaceous ant in mandarin orange trees to attack the insects which feed on the leaves. Farmers in China to this day still help the ants to travel from tree to tree by installing interconnecting bamboo rods as bridges.

Probably the first successful movement of a natural enemy

from one country to another occurred in 1762 when the mina bird was introduced from India to Mauritius to control the red locust, which was wreaking havoc there. It has proved to be very beneficial ever since, in particular in eating the cane borer—a pest of sugar plantations.

Erasmus Darwin (the grandfather of Charles) wrote in 1800, 'Cabbage caterpillars would increase their destructive numbers but one half of them are annually destroyed by a small ichneumon fly which deposits its own eggs on their backs'. Similar references to insects attacking others followed frequently through the century. Thus we find Sidney Oliff, an entomologist, writing in 1890:

> In the hop growing districts of the South of England swarms of ladybirds occur and I have myself seen them in such numbers that they had to be swept from the pathways. In seasons of scarcity [they] are collected and sold to the hop growers [of Kent and Surrey] who set them free, a practical application of one of nature's benefits which, as far as I am aware, is almost unique in the history of economic entomology, but one, nevertheless, that has prevailed for many years, if not for centuries.

The success of the ladybird

The first successful demonstration of the power of biological control on a large scale was carried out in California in 1888. The infant citrus industry developing in the wake of the gold rush was threatened with annihilation. A little insect, the cottony cushion scale (*Icerya*), was attacking citrus trees to such effect that farmers were abandoning their orchards. Something had to be done, and done quickly if the citrus industry was to survive.

At the instigation of Charles Riley, an entomologist, one of his assistants Albert Koebele was sent by the Fruit Growers' Association of California to Australia 'to seek out natural foes of *Icerya* and to bring them to California if any were found'. In Australia, Koebele went to work. Among other predators he noticed ladybirds feeding upon the eggs of *Icerya* in a little garden in Adelaide. In 1888 he sent 129 ladybirds to California; and they were bred on an *Icerya*-infected orange tree. The ladybirds (or vedalia beetles as they are sometimes known) destroyed nearly all the *Icerya*. They were then distributed to 288 different orchards with

$$\frac{A \mid B}{C \mid D}$$

1. Stages in the development of a Cabbage White butterfly, showing A eggs, B larva, C pupa and D adult (different magnifications)

2. Water hyacinth

3
―
4

3. Effect on
foliage of late
blight of potato

4. Modern
plough

7. Modern sprayer

7
——
8

8. Wheat field showing, on left, area treated with selective weedkiller; on right, unsprayed area

$$\frac{A}{B}$$

9. *A* Path infested with weeds; *B* same after treatment with simazine

10. Area of bracken showing a strip one year after spraying with asulam

11. Orchard spraying

12. *A* Dense prickly pear in Queensland, Australia; *B* same area after treatment with cactoblastis

similar successful results: they cleared the trees of the pest. The work was extended throughout California, and has remained one of the outstanding projects in biological control and an important milestone in applied entomology. A measure of its success is that, in a single year following treatment, more than double the number of carloads of citrus fruits were shipped from Southern California compared with the previous year. The ladybird continues to control the cottony cushion scale in California today as effectively as it did in 1888.

Koebele was fêted. In appreciation of his work, a fund was started and enough money was raised to present him with a gold watch and Mrs. Koebele with a pair of diamond earrings. An eminent entomologist later wrote (in 1925), 'I imagine that this presentation of diamond earrings to a wife of an entomologist as a result of his work in entomology is probably unique. I know of no other entomologist who has been able to see that his wife wears diamond earrings'.

It is not surprising that California became an active centre for biological control research, and successes there have certainly been striking. It has been calculated that control of the citrophilus mealybug, the black scale, the klamath weed, the grapeleaf skeletonizer, and the spotted alfalfa beetle by biological methods has led to a total saving by the citrus industry of California of $110m. from 1923 to 1958.

The prickly pear in Australia
Biological control can also be used against weeds; it has been outstandingly successful in the control of the prickly pear in Australia. The prickly pear (a native of Brazil) was introduced to Australia by the first colonists in 1788. It rapidly became established as an overgrown hedge. Other species of prickly pear were introduced in 1839 and in 1860 for hedging and grazing purposes. But prickly pears were not content to stick to the hedgerows, and one (*Opuntia stricta*) in particular was beginning to spread rapidly, enjoying the Australian soil and climate to such an extent that it was causing considerable concern. In fact by 1900 it had covered 10 million acres of land. In another 25 years it increased the total to 60 million acres and was spreading at the rate of one million acres per year. About half of the 60 million acres were occupied by dense prickly pear plants, growing to some 3–5 feet in height and completely covering the ground to

I

the exclusion of all grasses and other plants. The prickly pear had changed the face of the Australian countryside.

The land which it covered had been good sheep- and cattle-grazing ground, dairying country, and cereal- and cotton-growing soils. The value of the land at that time was estimated at between 5/- and 30/- an acre, but mechanical or chemical eradication could not be achieved for less than £10 per acre. Hence year by year more land was abandoned and homesteads given up.

In 1912 Queensland, worst-hit of all the states, initiated the investigation of insect predators and recommended the introduction of insects from America. The hunt was on. In 1920, explorers were sent to the United States, Mexico and Argentina to search for insects living on prickly pear. The answer came from Argentina. In 1925, 2,750 eggs of the Argentinian moth borer were sent to Queensland. The larvae hatching from the eggs were fed on the prickly pear, which they apparently relished. The stock was increased by breeding so that by the second generation 2½ million eggs were obtained; these were sent to twenty locations throughout the state.

By 1927 there were 9 million eggs available, and these too were distributed, but by this time the insect was becoming so common that it was no longer necessary to carry out a breeding programme. From 1927 to 1929 three thousand million eggs were made available from field collections. It was obvious that the moth borer was doing well, and as the moth thrived so the prickly pear suffered. Huge areas were being killed off at a time. In fact prickly pear became so scarce that millions upon millions of starving larvae of the moth borer began to die off too. And yet under such great stress the moth did not attack any other plant: apparently it could live only on prickly pear as a food source. And herein lay the great secret of success. It did not harm any other plants. Millions of acres formerly covered by prickly pear have now been returned to agricultural use. This work was an outstanding triumph in the biological control of weeds.

Control of St. John's wort
The control of the plant St. John's wort (so called because it flowers on or about 24th June—the feast of St. John the Baptist) is another very successful example of biological control. The weed is a native of Europe and was first reported in the United States in Pennsylvania in 1793. By 1900 it had spread to the

northern part of California around the Klamath river, where it became known as the klamath weed. By 1929 it had occupied over 100,000 acres of previously useful range land; and by 1940 it had extended to 250,000 acres.

It not only took over and crowded out useful range plants; it was also a danger to any cattle and sheep eating it, since the plant contains what is termed a photosensitizing agent. This substance when taken internally causes the animal's skin to become very sensitive to sunlight, resulting in severe blistering.

Two species of leaf-feeding beetles were brought over from Europe and released in the field in 1945; they quickly became established. One of them, *Chrysolina gemellata*, really attacked the weed. The adult beetles emerge from their pupae just below the surface of the soil in April and May and feed voraciously on the foliage. They then go into a resting stage in July and hide under stems or debris. Interestingly enough, the St. John's wort goes into a dormant stage at this time too. When the rains come in autumn, the plant begins to grow again, and again the beetles come out of their hiding places to feed on the foliage. Mating takes place in October, eggs are laid which develop into larvae which take care of any remaining growth from the autumn period. *Chrysolina* is a pretty formidable adversary since it feeds on nothing but the klamath weed. Many thousands of acres in California have now been returned to range land through its activities.

Greenhouses

Biological control is widely used in closed environments such as the greenhouse. In Britain it has been shown that a species of red spider mite (*Tetranychus urticae*), which is common on cucumbers, has been successfully controlled by a mite *Phytoseiulus persimilis* which feeds on it; the whitefly of cucumbers and tomatoes may be controlled by the introduction of the parasite *Encarsia formosa*; and other successes have been reported.

Fish

Fish have also been used to keep pests in control. Many of them feed on mosquito larvae and the use of this method is of long standing in malaria control. *Gambusia affinis*, the 'bogey of the mosquito nursery' has been introduced into several places in Europe. The fish can thrive in dirty water choked with weeds and also in water butts; an adult can eat one thousand mosquito larvae in a day. Fish have also been used to keep aquatic plants under control. Both the carp and the gouramy fish are greedy feeders on water weeds. And a sunfish introduced into Hawaii from South Africa in 1955 is playing a notable part in keeping the extensive ditch systems of that country free of weeds.

The Manatee

One of the most interesting animals used in aquatic weed control is not a fish, but a mammal. The manatee or sea cow, among the world's most harmless large mammals, has few enemies other than alligators and man, and is protected by law in many parts.

Singly or in groups of 15–20, manatees swim sluggishly. They are up to 15 feet long and 1,500 pounds in weight. There are three species: one is found off West Africa, one lives in the Caribbean from the south-eastern United States to northern South America and one thrives in the estuaries of the Orinoco and the Amazon. It is said that it was the sighting of manatees that gave rise to stories of mermaids: Christopher Columbus noted in his journal for January 1493 that he saw three mermaids off the coast of Haiti rising well out of the water. (Later he realized that they were manatees.) Manatees feed mainly at night, on aquatic plants, and they eat any vegetation within reach. During 1959–60 it was found that two manatees each $7\frac{1}{2}$ feet long were capable of cleaning a canal 22 feet wide and 1,000 yards long in seventeen weeks. They effectively keep reservoirs clear of a variety of aquatic plants, and have been distributed in several canals and irrigation schemes throughout British Guiana. It was hoped in 1960 that they might be used to clear the water of the hyacinth (see p. 15) that was spreading in the tropics of Africa and America and choking waterways, but there was only limited success because the manatees did not tolerate transportation, corraling or generally being kept in semi-domestication.

Rabbit control

One of the most striking examples of biological control was the myxomatosis epidemic which struck Britain in 1953. Myxomatosis is a virus disease native to South America—where it is not particularly virulent. It is passed from rabbit to rabbit by means of a mosquito in its native home. Many attempts to introduce the disease into Australia failed, but they eventually succeeded in 1950 to such an extent that the rabbit was virtually wiped out.

The disease was also deliberately introduced into France, and from there made its way to Britain. It spread rapidly through the country and by 1955 it had reduced the rabbit population by more than 90 per cent. In Britain it was spread by the rabbit flea, but it may also have been helped by man. It was outstandingly successful, but it is such a distressing and disgusting disease—producing swollen heads and open sores—that it was made illegal to spread it deliberately. A few hares became infected but otherwise all other mammals were immune; even infected corpses could be eaten by other animals without harmful effects.

The virtual disappearance of the rabbit had interesting ecological effects. An important food-source to carnivorous birds and mammals had disappeared. For two successive years buzzards, lacking food, failed to breed although there was no noticeable fall in buzzard numbers. Kites also failed to breed over the same period; and foxes turned more to hen-runs. But the most striking ecological effects were on the vegetation. Many interesting and beautiful plants flowered profusely, particularly in those areas that had formerly been hard grazed by rabbits. This was followed by the natural succession of scrubland so that many of these areas are now covered with shrubs and small trees.

In recent years both in Australia and Britain there has been some resurgence of the rabbit population. With the aid of other control measures, however, it looks as if the rabbit can be safely kept in check.

Finding a successful predator

It is interesting and perhaps salutary to note that despite its eradication in Australia prickly pear is a pest in other parts of the world—India, Sri Lanka, South Africa—and in these areas the moth borer has had little or no effect. It cannot be assumed that a predator successful in one part of the world will necessarily be successful in another. These and other considerations have led experienced workers to realize that it is not yet possible to predict how effectively an introduced predator will perform. This is something that can only be determined by trial and error: neither the behaviour of the predator in its native home nor its behaviour in the laboratory is a reliable index of how it will act in the field. Some would describe biological control as an art, not a science.

Nevertheless a list of certain characteristics has been compiled in an attempt to describe a potentially successful predator. These are: (1) it must have a high searching capacity, i.e. it must be able to go out and find the pest when it is scarce; (2) it should in general feed on only one species, the pest; (3) it should have a high reproduction rate so that numerous generations can be produced in a relatively short time; (4) it should thrive under the same conditions as the pest; and (5) it should breed readily in the laboratory so that initial distribution can be readily effected. This latter requirement does not apply to vertebrate predators.

Biological control involves potential dangers. There is the obvious risk that the predator, having disposed of the pest, will turn its attention in other directions and may indeed become a pest itself. There are several classic examples of disastrous results from introduced predators, but these have arisen because they were introduced before adequate tests were carried out. One such case concerns the mongoose in Hawaii. Hawaii had become infested with rats from the ships which docked there; finding a favourable climate, plenty of food, and few natural enemies the rats thrived. In 1883 it was suggested that the mongoose, a noted killer of rats, should be introduced from India. This was done, and the mongoose killed rats; but it also killed poultry, ducks, game and other birds. So now the Hawaiian Islands are left with a mongoose problem.

In some cases the food requirements of the predator are complex and must be met if it is to survive. For example, the grape-leaf hopper is a serious pest in California. A parasite—*Anagrus epos*—attacks the eggs of the leaf hopper and can effectively control the pest throughout the vine-growing season. But the parasite is only able to over-winter in the eggs of another harmless leaf hopper (*Dikrella cruentata*) which lives on wild blackberry bushes. Its continued existence therefore depends upon the continuing presence of blackberry bushes in the area. Where the vines and the blackberry bushes are widely separated, the efficiency of the predator is very much reduced. The solution to the problem obviously lies in ensuring that plenty of blackberry bushes are available.

The cost of biological control

One very appealing characteristic of biological control is that it is cheap. There may be quite a considerable outlay of money in the initial stages, such as in mounting explorations to search for suitable predators, and for growing the initial stock in the laboratory when found. Nevertheless, once the predator is established it requires little or no help thereafter.

Dr. F.J. Simmonds of the Commonwealth Institute of Biological Control recently calculated that a total expenditure of £1m. in seven successful biological control ventures had yielded aggregate benefits of £4,780,000 and subsequent annual savings of £262,000. Similarly Dr. P. de Bach estimated that work in the Department of Biological Control, University of California from 1923–59 cost £1,500,000 but led to a saving on five projects alone of £40,000,000 with a recurrent benefit of £10,500,000 a year. In the first year after the establishment of the myxomatosis virus of rabbits in Australia, the value of the wool clip was increased by at least £35,000,000. With biological control there is no residue hazard; and provided the predator sticks to its host, little if any risk is involved. Such control has always been of great value where profit margins are narrow and in areas which are not highly mechanized—which is not surprising. In addition, however, there have been cases of great success in areas of high development and technology, e.g. the California citrus area.

Microbial pesticides

Microbial pesticides may be described as those micro-organisms, or their products, that are capable of attacking and destroying pests. The use of such pesticides is in some ways really an aspect of biological control, although when one is dealing with microbial products it branches off into chemical control.

The concept goes back a long way to 1835, when Bassi demonstrated the infectious effect of a white fungus on a number of insects (including the silkworm); and in 1873 Le Conte put forward recommendations advocating the use of disease as a means of insect control. By 1888 spores of the fungus *Metarrhizium anisopliae* were being used in the field in tests against insect pests. In the ensuing years both successes and failures have been reported.

There have been some rather interesting developments of biological control by viruses in recent years. One is concerned with the control of the pine processionary caterpillar, which is a serious pest in the south of France. The caterpillars live together in webbed nests and they walk in procession, head to tail, which indicates that they could well transmit disease amongst themselves. It was found that they were susceptible to a certain virus which killed them off. The virus was therefore grown in living caterpillars in the laboratory by allowing them to ingest the virus with their food. The resultant dead caterpillars were next ground up with bentonite powder and then sprayed by air over an area of 350 hectares. This was done in 1958, and in a remarkably short time there was a 95 per cent mortality of processionary caterpillars, presumably due to caterpillars becoming infected and passing on the disease to others. Control was effective until about 1965 when there was some resurgence. It is considered that in order to be fully effective much greater areas need to be sprayed, and mass culture techniques are now under consideration. Virus diseases have also been used in Canada against the European pine sawfly, by spraying. In Australia a virus has been very effective in the control of the cabbage butterfly.

Among the bacteria, most promising results have been obtained using *Bacillus thuringiensis*. Field tests with this organism have indicated that it can be used for the control of many insects of field crops and forests. Another successful bacterium is *Bacillus popilliae* (the milky disease organism) which attacks the Japanese beetle, though results using fungi have been very variable.

There has in recent years been an increasing interest in substances formed by micro-organisms which are toxic to insects. In particular, the toxins produced by *Bacillus thuringiensis* have been the subject of much work. In all, some five have been isolated. Among the various types most interest has been centred on the 'parasporal body', a proteinaceous crystal produced inside the bacterium. The crystal dissolves in the alkaline midgut juices of the insect, affects the permeability of the midgut epithelium and enables the highly alkaline juice to permeate into the haemocoele (body cavity), making the liquid there more alkaline. This alkaline fluid may cause general paralysis followed by death. Only moth and butterfly larvae are susceptible to these toxic crystals. Relatively pure preparations of the toxic crystals can be obtained by several methods, but they are so chemically complex that it has not yet been possible to synthesize them in the laboratory.

A second toxin was discovered in a variety of *Bacillus thuringiensis* in 1959. This substance has been called the 'fly toxin' or

'Fly Factor', because it has been shown to interfere with pupation of the housefly. Apparently when the toxin is ingested it kills only flies and their relatives. Microbial pesticides have been applied with conventional dusters and sprayers and sometimes with mist blowers and from aeroplanes; they are dispersed by wind and rain and by carriers.

Predacious fungi

Among the most fascinating fungi are those that live by trapping and living on eelworms. These have been named the predacious fungi. There are two main groups: (1) eelworm trappers, and (2) internal parasites. The fungi of the first group capture eelworms alive by various devices including adhesive networks, adhesive knobs and, most interesting of all, by forming rings that constrict round the eelworm's body and hold it. Each ring is made up of three cells which are sensitive to touch; if an eelworm pushes its body into a ring, the three cells suddenly swell inwards, closing the ring and so gripping the victim tightly in a noose. This has been compared both with a rabbit snare and with the cowboy's lassoo. It appears to be at least equally efficient as either. The internal parasites, on the other hand, generally gain entry to the eelworm by attaching sticky spores to its body which then germinate and digest away the contents by means of the growing hyphae.

Despite many attempts, it has not been shown conclusively that any of these predacious fungi attack parasitic eelworms. They appear to prefer the free forms. However, some workers, including Dr. Duddington who has worked extensively in this field, believe that they have an important role to play in the biological control of eelworms.

The future of biological control

Biological control has chalked up a number of very impressive successes. In addition to those already mentioned, there are records of more than one hundred species of pests being controlled to a greater or lesser degree by natural enemies in more than sixty countries. Biological control has worked well in both temperate and tropical climates; it has been successful on islands and over continents, in northern and southern latitudes. There are some who believe that it is as yet only in its infancy.

One interesting suggestion by de Bach (one of the outstanding workers in this field) is that of importing what he terms an 'ecological homologue' for the competitive displacement of a pest. Such a homologue may or may not be a pest itself. For example, it is suggested that it might be possible to control species of biting flies that attack man by replacing them with species that would compete aggressively with the biting flies for food and thus eliminate them, but which had no capacity for attacking and biting man. In fact something like this has occurred in several parts of the world. When disease-carrying flies were reduced by various control measures, they were replaced and kept down subsequently by harmless species which were introduced to take their place.

Another technique which must have great potential in this field is the inducing of mutant species. All organisms have predators which attack them to a larger or smaller extent. In most cases, as mentioned earlier, a balance has been struck between host and predator: the host supplies enough food to keep the predator going, but the predator is not aggressive enough to fully overcome its host. This is the normal situation, but occasionally the predator may suddenly show a great increase in virulence and spread like wildfire (major flu outbreaks would be analogous situations). This is due to the fact that the predator has developed a mutant or 'sport' which differs from its fellows in its increased virulence. With the enormous range of mutagenic agents such as radiations and mustard gas that man has at his command he should be able to subject 'weak' predators to such mutagenic agents and, by careful selection, pick out and breed from those mutants which show an enhanced virulence. This would reduce the need to import predators from other countries.

Integrated biological and chemical control
One of the criticisms that advocates of biological control have levelled against pesticide users is that pesticides kill insects indiscriminately: as well as killing off harmful insects, a pesticide may also kill off beneficial ones such as predators. As a result of this, it is claimed that the use of pesticides may sometimes lead to a pest outbreak.

Pest outbreaks which occur as a result of pesticide usage may be due to the rise to economic importance of an insect relatively unaffected by the pesticide, while its normal predators—being very susceptible—are killed off. There was an example of this in

California when malathion, a drift contaminant, killed off the vedalia beetle which was very susceptible and thus allowed the resurgence of the cottony cushion scale (relatively resistant to malathion) which the vedalia beetle had formerly kept in check.

It is not surprising, then, that attempts have been and are being made towards a co-ordinated approach, combining the advantageous features of both chemical and biological control measures for reducing the pest while causing a minimum disruption of the natural enemy activity. Such approaches have been called 'integrated chemical and biological control'. This is an encouraging development, because it means that one group recognizes the importance of the activity of the other and that both have the same end in view—the suppression of pests and the consequent well-being of protected organisms.

All integrated control programmes have as their goal the preservation of natural enemy reservoirs, whilst at the same time destroying a satisfactory proportion of the pests. This can be brought about by (a) a differential exposure of the pest and predator to a given pesticide, or (b) by exploiting the difference in physiology of pest and predator. Regarding differential exposure it may be possible to retain reservoirs of predators by leaving adjoining fields of crops untreated. The reservoirs may also be laboratories where predators can be bred and subsequently released onto treated crops: this method has been used with some success against scale insects infesting Californian citrus trees.

A rather interesting form of selectivity makes use of the life histories of the organisms concerned. As we have noted in Chapter 1, many insects go through a number of developmental stages before they become adults. After hatching from the egg a larval stage is produced, followed by a pupal stage which then hatches to give rise to the adult insect. It has been found that the larval stages of many predators are tolerant to pesticides, while the pupal stages are quite remarkably resistant even to fumigants. Pest and predator do not always coincide at the various stages, so that by selective timing it should be possible to apply pesticide at a time when the pest is at its most susceptible stage and its predator at its most resistant.

Specialized feeding habits of predators are among the most distinctive of features which may confer on this group a degree of physical immunity to pesticide treatment. For example, if the predator is an internal parasite in the pest then it will obviously

127

be protected from a contact spray—although if it subsequently consumes all of the pest then it may also succumb to the effects of the poison.

The amount of pesticide applied is important. Recommendations for application are generally related to 'overkill': much more pesticide is applied than is required. By reducing the dosage it may be possible to exercise a 'selective' effect, either by preserving adequate predator reservoirs which would otherwise be destroyed or by permitting earlier predator re-entry after extermination. In either case, subsequent predator multiplication may more than offset any sacrifice in initial pest destruction through a lowered dosage.

Physiological specificity is obtained when a pesticide is more poisonous to a pest than to its predators. It is not, however, particularly common—many predators are very susceptible to pesticides. But selective toxicity has been shown for at least one systemic insecticide—schradan. Tests have shown that this insecticide is about three times as toxic to a pest such as the corn leaf hopper as it is to its predator, the mirid bug which lives on the eggs of the leaf hopper. A given schradan concentration has been shown to kill 98 per cent of leaf hoppers but only 20 per cent of mirids. It has been suggested that this selectivity is due to the ability of the mirids to convert schradan to a non-toxic compound.

Dung disposal in Australia
Biological control may sometimes be used against non-living things. An interesting project is at present being carried out in Australia, which is concerned with the introduction of beetles to aid in dung dispersal. In most countries cattle-dung pads break down and disperse in a matter of a few weeks at the most. This is because of the activity of numerous dung beetles. In Australia, however, the beetles are absent; the only native dung beetles are adept at disposing of the small pellets of marsupials but are incapable of attacking the cattle pads. As a result, dung pads may remain on the ground in Australia for five years or more. Since each beast produces ten pads per day there is a considerable problem. Cattle will not touch the rich green vegetation that develops round each pad, so that much of a pasture can rapidly become unproductive. In addition cattle-dung is the breeding-ground of two pests—the bloodsucking buffalo fly of cattle in the north, and the widespread and irritating bushfly.

Two types of beetle are being introduced, and results are promising. One type attacks maggots in dung pads; the other breaks up the pad and buries the dung in tunnels under the soil, thus aiding the fertility of the soil and at the same time hindering the development of maggots and parasitic worms. It is said that there are more than 1,000 species of dung beetle in Africa alone, and the Australians consider that it may be necessary to introduce a wide variety of them in order to effect successful and speedy dispersal.

The reader will have gathered from this chapter that biological control has been very successful in a number of cases. But not even its most ardent advocates would or should claim that it holds all of the answers. As Drs. D.F. Waterhouse and F. Wilson, two Australian workers active in the biological sphere have said:

> There have been a substantial number of spectacular biological control successes but there have also been many failures. Biological control does not meet all situations and cannot be regarded as a general alternative to pesticides. It can only be successful where it is possible to find a natural enemy that is powerful enough to overcome the pest or where we can develop an enemy powerful enough to do so. Compared with the number and extent of the problems that are involved the number of predators is small.

Perhaps biological control has in some ways been a victim of its own success: the outstanding successes of the ladybird and the moth borer perhaps led some people to expect too much. There have been subsequent successes, and there will continue to be so. But in the writer's opinion it seems clear that biological control can never be the main weapon in the attack on pests. It may be an important supplementary factor, and in some isolated instances it may even be the method of choice. However, the future may rather lie with *integrated* control, making use of both chemical and biological methods.

Chapter 10

NOVEL METHODS
OF CONTROL

As we have seen in previous chapters, the cornerstone of pest control is the use of pesticides, with some support being supplied by biological methods. However, within the past ten years or so scientists, as is their wont, have opened up new possibilities and have re-investigated old ones. Some of these methods have been shown to have practical applications; others are as yet untried on a field scale.

Sterilization
The use of sterile insects to control and eradicate insect populations is one of the most revolutionary ideas in entomology. The method consists of breeding insects in the laboratory, sterilizing them by chemicals or other means, and then releasing them so that these sterile insects compete for mates with the natural population. This reduces the reproductive potential of the natural population in proportion to the ratio of sterile to fertile insects present in the population after insects are released. If the ratio is 1:1 and the released sterile insects are fully competitive, the reproduction capacity of the natural population is reduced by 50 per cent; if the ratio is 9:1, the reproductive capacity of the natural population is reduced by 90 per cent.

The method has been used with great success against the screw-worm. Screw-worms are parasites of warm-blooded animals. If such animals happen to have open wounds then the female fly of the screw-worm lays her eggs in such wounds. The eggs hatch out within 24 hours, giving rise to larvae (the screw-worms) which feed on the living muscle tissues of the unfortu-

130

nate animal. The effect can be devastating, for screw-worms are voracious feeders. Small animals die fairly rapidly as a result of attack; and even a fully-grown steer can be killed by a heavy infestation in less than ten days. When the larvae finish feeding they drop from the wound and burrow into the soil to pupate. If the soil is cold the pupae die; if not then flies emerge from the pupae, mate when they are two days old, and in 6–8 days lay their eggs in wounds. Males mate repeatedly if they can find virgin flies. Females mate only once.

The screw-worm is limited in its year-round distribution to tropical and sub-tropical parts of the Western hemisphere. In the northern United States, the winter cold kills the screw-worm off. It can continue however in peninsular Florida, Southern Texas and Southern Arizona, where the temperatures even in mid-winter are sufficiently high. With the coming of spring the flies migrate northwards as far as New Jersey, Illinois and South Dakota.

The annual loss due to screw-worms in the USA was estimated in 1957 to be \$25–\$40m. Research was begun on the possibility of using sterile male flies. It was determined that both X-rays and gamma rays were equally effective in producing sterilization. The most efficient method was to irradiate pupae with a 7,500 Röntgen dose two days before the adult insect emerged. It was shown in laboratory experiments that radiation-induced sterility was permanent, and that the sterilized males were competitive with normal males in cage mating experiments.

Field studies on a major scale were undertaken on Curaçao, an isolated island of 170 square miles which swarmed with goats and sheep, most of which had screw-worm infestations. Screw-worms were reared in the laboratory, and the pupae irradiated in a cobalt-60 source at Oak Ridge, California. The irradiated pupae were packaged with a little excelsior in small paper bags at the rate of 130 per bag and transported by air to Curaçao, where the resultant flies were released over the island twice each week in flight-lanes half a mile apart. The release rate, after initial tests, was 400 males per square mile per week. The eradication of screw-worms from Curaçao was successfully accomplished by releasing these flies continuously from July 12th, 1954 to January 6th, 1955. All eggs laid were found to be sterile. The success on Curaçao was due in part to the fact that it was an island, an isolated community where fresh infestations could not take place.

The scientists now turned their attention to the south-eastern

United States, where the annual loss was calculated to be $10–$15m. This was a more difficult task. It was estimated that the release of four hundred males per square mile per week had been necessary to clear Curaçao, and that a similar rate of release would be needed in Florida. The over-wintering area for the screw-fly was calculated to be about 50,000 square miles, and this rate would require the release of twenty million males weekly. Since the sexes are equal in numbers and it was impracticable to separate them, this meant that 40 million flies had to be reared; allowing for mortality the figure rose to a massive 50 million flies per week! New rearing methods were developed, using thermostatically-controlled vats and feeding the insects on juice from frozen horsemeat. A radiation source was installed to sterilize 18,000 pupae every seven minutes.

In January 1958 releases of sterile flies at the rate of 1 million each week were begun, in an experimental barrier zone across peninsular Florida. A few screw-worms later appeared north of this line and were 'spot treated' by special releases of sterile flies. These procedures prevented a build-up of screw-worms and isolated the infested territory to the southernmost part of Florida. By 1959 the programme was successfully completed and Florida was free of screw-worms. In the summer of 1961 some made a reappearance, apparently introduced by shipments of infested cattle—in spite of inspection and insecticide spray stations at major highways crossing the Mississippi River. Cattle were therefore sprayed with insecticides and sterile flies were released locally, and the incipient outbreak was eliminated.

The success of the Florida experiment and similar experiments in Texas persuaded Congress to extend appropriate funds for screw-worm eradication to other parts of the United States; it is under way in the last remaining infested areas.

One major difficulty is the flight distance of the fly. Using marked flies it has been shown that they can migrate up to 300 miles, so that there is the constant threat of migration of flies from Mexico. At present the United States maintains a 1,200 mile lay barrier zone from Texas to California at a cost of about 5 million dollars per year. It is considered by some experts in the field that it might be more economic to eliminate the screw-worm from Central America, with a barrier at the much narrower isthmus of Panama.

As we have noted earlier, sterilization is readily effected by irradiating the pupae. Both X-rays and gamma rays are used because they penetrate tissues readily. Late in the pupal stage the insect is practically fully-formed and is ready to emerge—all the body cells are formed and mature. Both male and female reproductive cells are, however, in course of formation and as such are very susceptible to the rays. These rays act upon the chromosomes—the carriers of the hereditary messages—bringing about physical and chemical changes within them. These changes are known as mutations.

If the radiation dosage is low, only a few mutations will be brought about, and offspring resulting from these sex cells will be only slightly altered. If, however, the dose of radiation is increased then the number of changes in the chromosomes is also increased so that the offspring arising from sex cells containing such chromosomes cannot survive. Such changes are known as 'lethals'.

This is the principle on which sterilization works. The sex cells are not destroyed, but they are altered; they contain so many mutations that any offspring resulting from them will be dead. When an irradiated male insect mates with a normal female, then the irradiated sex cells (the sperm) are deposited in the vagina. The fertilized eggs which result proceed to divide, but some way along the developmental process the embryos die and the eggs do not hatch. It is therefore something of a misnomer to call irradiated males sterile since they do produce motile sperm; but since the embryos die due to this 'lethal' sperm, it is convenient to refer to such males as being sterile.

The same treatment that causes sterility in males causes sterility in females, but in most cases the females fail to produce sex cells and are thus more truly sterilized than the males. It may be noted that the body cells of both male and female insects are unaffected by dosages affecting the sex cells because the body cells have stopped dividing and are already mature before the radiation is applied. If the radiation is applied while the insect is at an early stage of development then the body cells would be similarly affected and the insect would die.

The demonstration of the successful use of the sterile fly technique in Curaçao stimulated research work on the use of this technique against a variety of other flies. Dr. Knipling, a pioneer in the field has listed five conditions which should prevail for the technique to be successful. These are:

K

(1) A method of mass rearing of the insect must be available.

(2) Adequate dispersal of the released sterile males must be obtained.

(3) The procedure of sterilization must not adversely affect the mating behaviour of males.

(4) The female of the insect to be controlled must normally mate only once, or if more frequent matings occur the sperms from irradiated males must compete with those from fertile males.

(5) The population density of the insect must be inherently low or the population must be reduced by other means to a level which will make it economically feasible to release a dominant population of sterile males over an extended period of time.

Using these criteria the technique has been applied to a number of flies. Tests were carried out against the Mediterranean fruit fly on the island of Hawaii in 1959–60. By releasing some 18·7 million treated flies at weekly intervals in an isolated mountainous area, the wild population was reduced by 90 per cent by August 1960; but re-infestations from outside the area soon brought the test area back to its normal level.

Greater success was achieved on Guam against the oriental fruit fly in 1963. Two typhoons had destroyed most of the tree-fruit hosts, thus markedly reducing the fruit fly population. Seventeen million sterile flies were released at the north end of the island where some fruit fly hosts had survived the typhoon, and the oriental fruit fly was eliminated. Eradication of the melon fly was similarly achieved on the island of Rota in 1963, by first reducing the melon fly population nearly 75 per cent by spraying with a protein bait containing malathion and then releasing some 280 million irradiated adults.

The effect of radiation on over half a dozen species of mosquitoes has also been studied. The experiments cannot be regarded as highly successful, but much useful information has been obtained which may be applied in the future.

Dr. R.C. Bushland, an active worker in the field, considers that despite some successes, our knowledge of the technique is still in its infancy; and that we are merely on the threshhold of realizing the full potential of this approach to insect control.

Chemical sterilization

In recent years, chemosterilants have received considerable

attention because of the possibility of using them to replace X-ray and gamma ray irradiation for the sterilization process. A wide range of chemicals can be used, and it is of interest to note that the search for effective compounds mirrors the progress in the development of cancer therapeutic agents. Since the object in both cases is the same, namely the prevention of cell division, it is not surprising that anti-cancer chemicals have been used as sterilizing agents; and conversely it has been suggested that insects may be very suitable test material for the testing of potential anti-cancer drugs.

Among the chemicals that have been used are the nitrogen mustards, folic acid analogues, and colchicine which is a natural product extracted from the autumn crocus. A substance 2,6-dimethyl-p-hydroquinone, extracted from the garden pea, has also been shown to be a potent sterilizing agent.

How the chemicals work

When used on females, no eggs are produced or the fertilized egg does not develop. The most commonly-observed effect is a cessation in the development of the ovaries, which may regress and disappear. Occasionally the ovaries may appear to be unaffected but the treated females do not produce any eggs.

In contrast we find that in treated males the development of the testes is seldom interfered with, but that the sperm which are produced—although appearing normal—do not produce any viable offspring. Presumably they contain lethal mutations.

Application of chemicals

Chemosterilants could be applied as dusts or sprays in the same way that insecticides are used now. This would be especially useful for ponds, ditches, rivers and so on but much detailed research will have to be done into side-effects on other organisms before such powerful biologically active compounds could be used in such an uncontrolled manner.

Similar remarks apply to other methods that have been suggested, such as using baits which have been treated with chemosterilants. It has been shown for example that cornmeal baits are useful as carriers. In an isolated refuse dump in Florida, research workers obtained excellent control of houseflies by using such baits containing the chemosterilant known as aphoxide. The same workers sharply reduced the housefly population in a poultry house by applying such baits to the poultry droppings. It is obvious then that such techniques could be used on a wide scale, and no doubt they will be.

Some rather ingenious methods have been tried out. In trials with the Mexican fruit fly, the chemical sterilization of the flies was accomplished by dipping the pupae of artificially-reared flies in the chemical for a few seconds and then allowing them to dry. A thin sticky layer of chemical remained on the pupal case. When the flies hatched out they walked over the pupal case and in so doing they picked up enough of the chemical to render themselves sterile. The trials were very successful, but reinfestation occurred.

In other experiments, bait containing protein food particularly attractive to the flies was hung in traps in the trees of a mango orchard. The flies, attracted by the smell of the protein, entered the trap, within which there was a small reservoir of water containing the sterilant. Flies drinking the water were rendered sterile and a very substantial reduction of infected fruit was subsequently obtained.

As yet, however, the sterile male technique, which has met with such success when radiation is used, seems likely to be the main solution. There is no residue problem and no risk of affecting other organisms in the environment, since the chemicals are applied in the laboratory. They may be administered in one of three ways:

(1) by mouth—by mixing the chemical in with food or drinking water

(2) topically—by applying to the outside of the insect so that it is subsequently absorbed through the cuticle

(3) by injection—though this is time-consuming and difficult.

But it is not only against insects that chemosterilants may be used. Many vertebrates including birds and mammals are pests. The possibility of controlling the reproduction of such pests would seem to be a viable one. Although they are pests, many of them are also rather attractive creatures; and, provided that they could be kept in check and their numbers kept low enough to pose no real threat to man, one would not want them eliminated.

In the search for anti-fertility agents, most of the research has concentrated on the ovum; but this concentration will not long remain. The sperm must be just as vulnerable to chemicals as is the ovum; and since in nature many females may be serviced by one male, it would seem to be more economical to attack one male rather than numerous females. It is of interest to note that Dr. W.H. Elder conducted a four year search in the United States for a practical oral contraceptive for controlling nuisance birds such as pigeons; but most substances that were effective in inhibiting ovulation in other animals proved to have little effect on pigeons, even in nearly lethal doses. Among the chemicals he tried out were tranquillizers, gametocides, anti-thyroid compounds, fungicides and insecticides.

On the other hand, a number of compounds have been found to be active against male birds. In investigating anti-fertility agents, account must be taken of whether treated birds or animals lose their aggressiveness as a result of treatment. Should this happen and they become submissive, they could fail in the quest for mates and thus the whole purpose of the treatment would be lost.

The search for mammalian anti-fertility agents is a particularly active one. Both oestrogens (female hormones) and androgens (male hormones) have been shown to be very potent. Small amounts of either were injected into new-born rats, producing long-term effects: as the rats grew older it was found that they were incapable of producing either sperm or ova; and this effect was permanent. The chemicals apparently acted by affecting the growth of the pituitary gland which controls sexual development.

However, not all infertility agents are permanent in their effects. With some the effects wear off and the treated animal later becomes as fertile as it was originally. Nevertheless such agents can be very important weapons, particularly against animals that produce only one litter per year. The chemical agent can be mixed with a bait and made available at the right time of year. The temporary sterility agent would also probably find more favour on ethical grounds too, simply because its effect is not permanent.

Sex lures

One of the most interesting methods of control that has been developed in recent years is the use of sex lures. The virgin females of some insects emit substances attractive to the males of their own species. These chemicals are known as pheromones. Very small amounts of some of these substances are effective. It has been found that the female gypsy moth can attract males from more than two miles away. This must mean that her chemical attractant is active at the molecular level—that a few molecules may be enough to excite and direct the male to his mate. This speaks much not only for the potency of the chemical but also for the male's receiving apparatus, the antennae of the male gypsy moth being complex, highly-developed organs.

Other insects, while not having the 'distance' effect of the gypsy moth, make up for this in numbers; for example, one oak eggar female in one experiment attracted more than 60 males over a period of fifteen hours. Females do not, however, have a complete monopoly over sex attractants. The males of the boll weevil can release a chemical that draws females from some considerable distance.

It is not surprising that sex attractant chemicals have been the subject of investigation—not only are they of interest in the study of insect physiology and behaviour, but they could also be used for insect control. To date a number of these chemicals have been identified, thanks to the development of very sophisticated micro-chemical techniques. Most of them are very complex organic substances; for example that of the silk worm is *trans*-1-*cis*-hexadecenol, that of the gypsy moth 10-acetoxy-*cis*-7-hexadecenol and the honeybee attractant is 9-keto-*trans*-2-decanoic acid. And there are others.

Once a sex attractant has been identified, the chemists get to work to synthesize related compounds with similar or even enhanced activity. One compound that has been synthesized is 'Gyplure', which, although it is not as potent as the original gypsy moth attractant, is much less costly to prepare. In addition, a number of compounds which are not even remotely related to the natural substances have been found to be very active attractants. For example caproic acid is a powerful sex lure for adult males of the Pacific coast wireworms; the lure causes the males to assume the mating position and attempt copulation.

The chemical released by male bark beetles is simultaneously a sex attractant and a food attractant. Certain males of the bark beetle initially find and attack a suitable tree, burrow into it and feed on the inner tissues. After feeding they release faecal pellets which contain a chemical attracting both male and female bark beetles to the vicinity. The females are stimulated to enter the galleries formed by the males which have released the attractant, whilst the new males construct new galleries when they encounter the tree. This is as astonishingly successful survival mechanism, since not only is mating effected but a suitable food source is also provided.

Food lures

In general, food lures are neither as potent nor as specific as sex lures. Fermenting sugar solutions and digested protein substances have been widely used for a number of insect species with varying degrees of success. They usually attract both sexes, but there is an interesting exception in the case of the mosquito where carbon dioxide attracts the females but repels the males. The reason for this is not known, but it is known that a substance that is a repellent for one species may act as an attractant for another. For example, of the many materials tried as mosquito repellents, two of the successful ones are kerosene and oil of citronella. It is therefore interesting that kerosene is an attractant for the Mediterranean fruit fly, while oil of citronella is a powerful lure for the oriental fruit fly. Some food lures may be made more potent by adding chemicals to them; for example, fermenting substances are apparently much more attractive to insects when a few drops of pyridine are added. It is of interest to note here that humans secrete lactic acid in sweat and on skin, and that this is a powerful attractant for female yellow-fever mosquitoes (*Aedes aegypti*). In experiments which have been carried out, a good correlation was found between the attractiveness of an individual to mosquitoes and the quantity of lactic acid present in an acetone washing of his hands. It is therefore possible that lactic acid might prove to be a valuable attractant in traps for the yellow-fever mosquito.

Oviposition lures

Closely related to food lures are substances that attract gravid female insects and induce them to lay eggs. For example, ammonia has a strong attraction for gravid houseflies and ammonium carbonate is a powerful oviposition lure for the greenbottle fly. Petroleum oils are reported to have the power of attracting biting midges.

Repellents

Some of the oldest methods of repelling insects are still in use today—fly swatters, feather dusters, fans and even horses' tails

139

were amongst the eàrliest physical methods and are still popular in many tropical and sub-tropical countries.

The first chemical method of repelling insects no doubt came with the discovery of fire, when it would be found that smoke, especially from wood fires, was very effective in dispelling and discouraging insect populations. From smoke in general to smoke from particular materials was but a short step; the efficacy of pungent odours would also no doubt be quickly discovered and strong-smelling substances hung in the cave or house, or perhaps even worn as a charm-bag round the neck.

This led in time to the identification of certain constituents and so to such substances as oil of citronella and oil of camphor. Oil of citronella, which is extracted from the plant *Andropogon nardus*, was the most widely-used mosquito repellent until 1938 and was the standard against which new repellents were measured. Bordeaux mixture (which we have already noticed—see Chapter 6) was one of the first fungicides, and is often cited as one of the earliest repellents used in agriculture after it was noted that insects usually avoided sprayed plants.

There are two classes of repellents: (a) physical, and (b) chemical. Physical repellents include dust, granules, water, oils, and waxes which may occur naturally in the surface of the plant or animal being attacked or may be applied to such surfaces in order to keep the attacker off. Auditory repellents are also included in this class. Amplified sound has been reported to be effective in repelling moths, sand fleas, various species of mosquitoes, and starlings.

Chemical repellents may be of either the vapour or the contact type. The former are probably more common. Not only must they be obnoxious to the insect and thus drive it off but they must at the same time be acceptable to the host, particularly if the host is man, so that no discomfort is caused. Research in chemical repellents over the past 30 years or more has been largely directed towards the production of synthetic chemicals for the protection of man. The requirements of the Second World War brought about a major programme of screening for such compounds, particularly for those that could be applied to the skin under tropical conditions. By the year 1945 some 7,000 compounds had been received and tested by the Entomological Research Branch of the United States Department of Agriculture; between 1945 and 1952 a further 4,000 were tested. At first the chemicals were tested for insecticidal and repellent action against body lice, mosquitoes and chiggers. In the later period tests against horse-flies, ticks and fleas were included. Of all the chemicals tested, four were recommended for skin application and two for clothing treatment. Since 1952 many others have been tested and several are now on the market. The most effective repellent appears to be diethyltoluamide, which is active against a wide range of insects including mosquitoes, flies, chiggers and biting flies. Recently the emphasis has been on new formulations to make the repellents more cosmetically acceptable and easier to apply.

One particularly interesting area of research is the hunt for an oral repellent, i.e. one which can be taken by mouth and which will then provide to be a repellent for mosquitoes and an aid in controlling itching. But such hopes have not been realized. Many insects, millipedes and others have their own repellents which they use for defensive purposes. Some cockroaches, earwigs, stick insects, stink-bugs, caterpillars, scorpions and millipedes actually discharge their repellents as a spray, sometimes to a distance of several feet; their aim is often unerring. In the whip scorpion, for example, two glands open at the tip of a short re-volvable knob that acts like a gun turret.

Several groups of research workers around the world are now engaged in elucidating the nature of the various secretions, and dozens of active principles have been isolated and identified. Most of them are of low molecular weight and represent com-pounds previously well-known in other connections such as hydrogen cyanide, acetic acid, caprylic acid and formic acid. They are used against a variety of predators including ants, praying mantids, carabid beetles, spiders, toads, lizards, jays, armadillos and mice.

Antifeedants

Antifeedants are substances which make the crop plant distaste-ful to the insect attempting to feed. It may take a bite, then wan-der off to another part of the plant for another try. Finding this also distasteful it eventually gives up and may die of starvation or move on to another untreated plant. If the untreated plant could be a weed and the antifeedant is properly formulated, then not only would the crop be protected but weeds might also be des-troyed. This would of course be ideal, but none of the anti-feedants so far treated has shown sufficient activity to be of commercial interest. The first antifeedant used in agriculture was ZIP, which is still used to keep rodents and deer from feeding on the bark of trees; but it cannot be applied to the leaves of trees because it destroys them.

Electromagnetic energy

Various forms of electromagnetic energy have long been recognized as having possible uses for insect control. These include all kinds of radiations—radio frequency, infra-red, visible light, ultra-violet, X-rays and gamma rays. Various electron tube oscillators are available thanks to the development of radar, radio communication etc., and these have been used in the medical field for many years for therapeutic heating of the body tissues in diathermy. Thought has now been given to using them for insect control, but so much energy is needed for a range of even a few feet that large-scale control by this method has been abandoned as impractical. The method has, however, been used experimentally for the destruction of stored grain insects such as weevils and beetles; but no practical scale applications of significance have been developed.

Direct application of infra-red radiation to infested products has resulted in the destruction of insect species. But the cost is high, and the method has not been adopted in practice. There are, however, some interesting experiments being conducted into the use of infra-red rays as an attracting agent. It has been shown that mosquitoes are attracted to insect traps containing lamps emitting infra-red light. It has been suggested that the ability of many male moths to locate females at a great distance may involve infra-red radiation from the female. Some workers have commented on the similarities between certain radar antennae and the structure of the antennae of moths, and that the spacing of the pectinations on such antennae would indicate an operating wavelength in the infra-red region.

Insects respond to visible and ultra-violet (UV) radiation in many ways. The most important from the point of view of control is that in general they are attracted to such light. Incidentally 'vision' for insects extends well into the near ultra-violet region, i.e. their eyes are much more sensitive to shorter wavelengths than are those of man. Laboratory studies have shown that large amounts of ultra-violet in the light output of lamps adds to their attractiveness.

Many types of trap have been devised to capture insects attracted by light, but basically all consist of an attractant lamp and a collection device which guides the insects into a collection can containing a volatile killing agent. Such traps are widely used in survey work, for determining the spread of a particular pest and predicting the need for control measures. Their use for direct control has been somewhat reduced by the introduction of pesticides, but since the cost of traps is low there may be some revival of interest in them on a local scale.

Ionizing radiations
Ionizing radiations have been used extensively for the control of insects, particularly in stored grains. X-rays, gamma rays and beta rays are all effective, but X-rays are costly to produce compared with gamma rays which can be readily obtained from certain radioisotopes. Beta rays, which are energetic electrons emitted from nuclei of disintegrating particles or electrons artificially accelerated to very high energies, are also readily obtained but are not as penetrating as gamma rays. The lethal and sterilizing effects of all three are about equivalent in their action on insects. In general the mortality of insects exposed to radiation increases with increased dosage, and complete mortality occurs in a shorter time as the dose is increased.

For any ionizing treatment to be effective it must kill off or sterilize the insects without harming the grain. Fortunately, this is in general the case—although radiation treatment cannot be used in grain that will subsequently be used for seed purposes. This is because not only does it stunt subsequent growth but it may also bring about many mutations.

A great many studies have shown that radiations used to control insects have no harmful effects on the milling and baking properties of the cereal grains so exposed. It may be noted here, however, that much higher energy levels are required to achieve complete sterilization of foodstuffs against bacteria, and these can produce off-flavours. Regarding the effect of radiation on the nutrient value of irradiated foods, tests have shown that no toxic substances are developed. Fat-soluble vitamins are quite easily destroyed by radiation; but water-soluble vitamins are no more affected than by normal cooking procedures.

Novel Methods of Control

Insect hormones as insecticides

A possible additional method of insect control has arisen recently, as a result of studies of the life histories of insects. As we have already seen, some insects go through the stages of larva and pupa before reaching the adult stage. The development of these various stages is controlled by certain chemicals, termed hormones, which are produced within the body at the appropriate stage. For instance, there is the 'brain hormone', which acts on certain glands causing them to release a second hormone, 'ecdysone', which brings about the moulting process. One hormone in particular which seems to play a vital part has been termed the 'juvenile hormone' or neotenin. This hormone is synthesized by two tiny glands in the head called the *corpora allata*, which also control the flow of the hormone into the blood. At certain stages this hormone must be present and at others it must be absent, or further development will not take place; thus, it must be absent from insect eggs for the eggs to undergo normal development. An immature larva has on the other hand an absolute requirement for juvenile hormone, or it will not progress through the usual larval stages. Then, in order for a mature larva to change into a sexually mature adult, the flow of hormone must stop. Still later, after the adult has fully formed, juvenile hormone must again be secreted. It can therefore readily be seen that this hormone is the key to development.

The hormone was first extracted, using ether, as a golden oil from silkworm moths in 1956. Its chemical structure has been determined. Subsequently juvenile hormonal activity has been found in extracts of the tissues of many invertebrates and vertebrates, and in plants and micro-organisms. These results indicate that we are dealing here with a group of substances which are widespread in nature.

Tests using the pure hormone as a spray have shown that it has devastating effects on numerous insects, e.g. those carried out on mealworm indicated that one gram of the hormone could kill about one thousand million insects. Chemists have set to work to synthesize similar materials in their laboratories. One of these will kill all kinds of insects and it has been calculated that 10–100 gm of this material would be sufficient to clear all insects from $2\frac{1}{2}$ acres. These various materials are selective in the sense that they will only kill insects and have no effects on other forms of life. It should therefore be possible to kill insects without harming other plants or animals, including man. Further, insects will not find it easy to evolve resistance to their own hormone.

But there is a problem as yet unsolved: hormones do not discriminate between the relatively few insects that are pests and the vast majority that are either helpful or at worst do no harm. The real need is for a hormone pesticide to kill only pests, and there have been interesting leads in this area. The following story also illustrates how chance can play a major part in scientific discoveries. A Dr. Slama from Czechoslovakia arrived at Harvard to work with Dr. C.M. Williams (a very distinguished worker in the field of insect hormones), bringing with him some specimens of the European bug—a species that he had been rearing in his laboratory in Prague for ten years. At Harvard he found to his distress that the bugs invariably did not reach sexual maturity. Instead, they continued to grow as larvae or moulted into adult forms that had many larval characteristics. It began to look as if they had access to a continuing supply of juvenile hormone. Eventually the source was traced to the paper towelling that had been placed in the rearing trays! Almost any paper of American origin had the same effect, whereas paper of European or Japanese manufacture had no effect.

The main source of pulp for paper in Canada and the northern USA is the balsam fir, and this tree was indeed found to have a juvenile hormone which Williams and Slama named the 'paper factor'. The active ingredient has been isolated and identified; it is close in chemical composition to the other juvenile hormone but differs in its action in that it only appears to be active against certain kinds of insects. It is fascinating to find this substance in a tree, and one can only conjecture that it has some protective function which has been the result of selection through the millions of years of evolution.

Chapter 11

THE YIELD

Introduction

For thousands of years, man was more or less powerless against his pests and was almost completely at their mercy. Knowing nothing of entomology or microbiology he turned to mystical explanations for the disasters that befell him, and the onslaught of a pest was sometimes looked upon as a God-sent punishment for his misdeeds. This view was supported by priests because it increased their hold over the members of their tribe. Sometimes a pestilence could also be used to benefit the people—thus Jehovah unleashed a plague of locusts to persuade Pharaoh to free the Children of Israel.

Dr. N.E. Borlaug describes the scene in his McDougall Memorial Lecture given at the FAO Conference in November 1971.

During the long, obscure, dimly defined pre-historic period when man lived as a wandering hunter and food gatherer chronic food shortages must have forced man to expend virtually all of his energies in struggling to feed himself. Under these conditions the growth of human population was also automatically slow because of the limitations of his food supply. In the misty, hazy, past as the Mesolithic Age gave way to the Neolithic there suddenly appeared in widely separated geographic areas the most highly successful group of inventors and revolutionaries the world has ever known. This group of neolithic men and women domesticated all the major cereals, legumes and root crops, as well as all the most important ani-

146

mals that to this day remain man's principal source of food. Some 9,000 years ago man had become a farmer. But this did not permanently emancipate man from the fear of food shortages, hunger and famine. Droughts, outbreaks of disease and insect pests, ravaged crops and flocks and famine resulted— they were recurrent catastrophes during the ages and Man's abilities to prevent them were limited.

Population

The present world population is three thousand five hundred million people; this is roughly 4–5 per cent of all those who have ever lived. In *Population, Resources, Environment* Paul and Anne Ehrlich list some interesting figures and put forward some fascinating projections. They estimate that around 8000 B.C., at the dawn of agriculture, the world's population was five million (the present-day population of Scotland). By the time of Christ it had increased to 200–300 million (the present-day population of the USA), with the development of agriculture. By 1650 it had increased to 500 million: i.e. it had taken 1,500 years for the population to double itself. By 1850 it had doubled again to 1,000 million, in 200 years. It took only a further 80 years (1850–1930) to double again to 2,000 million. The present rate of increase is 2 per cent, so the next doubling will take place in 35 years. Incidentally, if this rate of growth continued for the next two thousand years the world's population would exceed a billion billion (using the American notation)—some 1,700 persons per square yard of the Earth's surface, land and sea! It is of interest to note the trends in the various regions of the world (see Table 1, p. 148). In North America, Canada and the USA there is a growth rate of 1·1 per cent per year, which means a doubling of the population in 63 years. Latin America has the highest growth-rate of any area—2·9 per cent—which means a doubling of population in 24 years. In Europe the growth-rate is relatively low—0·8 per cent—but even this gives a doubling time of 88 years. (The United Kingdom's doubling time is 117 years.) The African growth-rate is 2·4 per cent and its doubling time 28 years; the pattern here rather resembles that of South America. Asia's current growth-rate is 2 per cent and the doubling time 35 years. It should be noted though that Japan is unique in this sector, her growth-rate being 1·1 per cent (similar to that of North America).

TABLE I

Region	Present population (millions)	Growth rate	Time to double (years)
North America	225	1·1	63
Latin America	276	2·9	24
Europe (excluding European USSR & Turkey)	456	0·8	88
USSR	241	1·0	70
Africa	344	2·4	28
Asia (excluding Asiatic USSR)	1990	2·0	35

(After Ehrlich and Ehrlich, *Population, Resources, Environment*.)

The food requirement

As the Ehrlichs point out, the story of human population growth is not a story of changes in the birth rate but of changes in the death rate. Famine has been an important periodic contributor to high death rates. More than 200 in Great Britain alone were recorded between A.D. 10 and 1846. It has been calculated that there were 1,828 Chinese famines in the 2,000 years preceding 1911— a rate of almost one per year; some of these resulted in millions of deaths. Even in this century famine has killed millions. Five to ten million deaths have been attributed to starvation in Russia (1918–22, 1932–4), as many as four million deaths in China (1920–1) and 2–4 million deaths in West Bengal (1943).

The Food and Agricultural Organization of the United Nations foresees that by the year 2000—one generation hence—overall food supplies will have to be expanded by 160 per cent for Africa, 240 per cent for Latin America and 200 per cent for East Asia, merely to provide a minimum adequate diet for their peoples.

Population growth in the better-off countries over the next 30 years is projected at almost 500 million (about 190,000 additional beings per day). By present standards each of these will consume directly or indirectly six times as much green plant tissue as that eaten by an individual in a poor country. The actual effect of 500

148

million new affluent people on earth is therefore approximately equal to the impact of the 2·8 thousand million additional human beings expected in the developing countries by the end of this century.

In the world today we have, averaged out, 12·5 acres per person. No more than three of these acres could be cultivated. By the year 2000 the 12·5 acres will be reduced to 1·5 acres gross per man, woman or child—counting Antarctica, the Sahara, and Mount Everest. Tillable land per person will be less than one half-acre.

The toll taken by pests
The toll taken of food by pests is high (see Table 2). In addition, the Indian Food and Agriculture Ministry estimated that in 1968 rats devoured almost 10 per cent of India's grain production (it would take a train almost 3,000 miles long to haul the grain India's rats eat in a single year). And yet, in 1968, India spent about 800 times as much on importing fertilisers as was spent on rat control. In one year rats in two provinces in the Philippines

TABLE 2

Region	Pests (*Insects*)	Loss per cent through		
		Diseases	*Weeds*	*Total*
North & Central America	9·4	11·3	8·0	28·7
South America	10·0	15·2	7·8	33·0
Europe	5·1	13·1	6·8	25·0
Africa	13·0	12·9	15·7	41·6
Asia	20·7	11·3	11·3	43·3
Oceania	7·0	12·6	8·3	27·9
USSR and People's Republic of China	10·5	9·1	10·1	29·7

(From *Plant Protection and Well-being*, H. Speich, CIBA-Geigy Journal No. 4, Winter 1971/72.)

L

devoured 90 per cent of the rice, 20–80 per cent of the maize and more than 50 per cent of the sugar-cane. In Africa birds destroy each year crops worth more than $7m. Insects and moulds damage a large amount of food in storage. Broadbent has estimated that total loss in store and in transit may be 10–15 per cent.

As well as hunger, disease has also through the ages taken a heavy toll of human life. Often the two went hand in hand and were until comparatively recent times together responsible for keeping the human population in check. Due to the onslaught of the potato blight in Ireland in the 1840s, millions died of the combined effects of hunger and disease. Typhoid and typhus were the constant companions of the Irish as they fled from their stricken land to Britain, the USA and Canada.

We have noted in Chapter 1 how plague often struck with terrifying ferocity, sometimes threatening the very continuation of the human race; of how typhus swept in epidemic form through towns, countries and continents; and of how the greatest killer of all—malaria—made large tracts of land unfit for human habitation, killed millions and rendered the lives of many millions more little other than a wearisome burden—and not a very long one at that. But these are just a few of the insect-borne diseases that have threatened man's existence, and which would continue to do so were the insects—the vectors of the diseases—not held in check by pesticides.

Changes brought about by the use of pesticides
The advent of pesticides is changing the picture in both agricultural and medical areas in many parts of the world. People need no longer face famine when their crops are threatened by disease or insect attack. Even the once-dreaded locust, so swift and omnivorous in its attacks, can be halted by the use of insecticides such as gamma-BHC; the potato blight will never strike again as it did in Ireland in the 1840s, now that we have foliage protectants; and malaria and typhus, two of the great killers in history, are being contained by the use of pesticides.

Millions were probably spared the agony of typhus after the Second World War by being dusted with DDT powder. Almost 80 per cent of that half of the world's population which twenty years ago lived in malaria-infested areas, is now protected, not to mention the millions who no longer succumb to plague or yellow fever. DDT in particular and pesticides in general are indispensable because none of the other methods available can work alone on the scale required.

Agricultural gains

The enormous gains derived from pesticides may be illustrated by work quoted by W.B. Ennis in a paper delivered at an FAO symposium on crop losses. On 8,100 hectares of heavily weed-infested rice in five states in the USA, the use of the herbicide propanil increased average yields by 50 per cent. The range was from 36–74 per cent. On less weedy fields the yields were increased by 40 per cent. In experiments carried out in Peru, it was shown that weeds reduced the rice yield by 34–68 per cent; and in the Philippines the use of propanil increased rice yields by 43 per cent. In the USSR, the yield of spring wheat in certain areas was increased 65 per cent by the use of the selective weedkiller 2,4-D. In Latin America, maize yields were reduced by 53 per cent by allowing weeds to grow; and in the USA yields of sugar from cane were reduced by 76, 52 and 42 per cent when weeds were permitted to compete with sugar-cane for 3, 6 or 9 weeks respectively after planting.

Weeds can of course be reduced by manual labour, and where this is plentiful and cheap then there is no problem. But in all countries of the world labour is becoming more and more expensive as people (rightly) demand higher standards of living. There is also a movement of people away from the land; and the land area for growing of crops is being reduced as more and more is being used for industrial development, roads and houses. For example, in the fourteen years from 1950 till 1964 the population of the UK increased by 8 million. In the same period $1\frac{1}{2}$ million acres of agricultural land were lost to urban and industrial development and the total number of people employed on the land fell by more than 220,000 (from 806,379 to 585,800)—and yet the total volume of agricultural output is nearly double what it was in 1940.

An authoritative book giving a clear answer to Rachel Carson's *Silent Spring* was written by J.L. Whitten in 1966. Mr. Whitten is a Congressman and Chairman of the House Appropriations Sub-committee of Agriculture. Recognizing the need for an objective study of the overall pesticide question, he set up a committee in 1964 to 'conduct an enquiry into the effects, uses, control and research of agricultural pesticides as well as an inquiry into the accuracy of the more publicized books and articles which increase public concern over the effects of agricultural pesticides on public health'. Over 185 outstanding scientists and 23 physicians were interviewed, as well as officials of the American Medical Association and University Medical School faculties. Included were biochemists, biologists, chemists, entomologists, nutritionists, pharmacologists, plant pathologists, toxicologists, zoologists and geneticists as well as experts on agriculture, conservation and public health. Whitten's book provides the results of that investigation. There is a great need for more facts and less emotion in the pesticide field; and many of the agricultural facts which follow have been taken from Mr. Whitten's book.

In the matter of quality and quantity of food the Western world is the envy of the rest of the world, and much of the credit goes to pesticides—even the hungry world is a little less hungry as a result of their use. As we have seen, practically every crop has its insect pests; and in some cases where the pest has gained the upper hand the crop has been practically eliminated. Attacks on apples by the codling moth in the States have led to the abandonment of many orchards, and yet tests have shown that had pesticides been available this situation could have been avoided. In trials in West Virginia, apple trees that had no sprays after the petals of the apple blossoms dropped showed 87 worms of the codling moth per 100 apples—whereas on trees sprayed with the insecticide azinphos-methyl, 100 apples showed only four worms. The corn earworm is a little green caterpillar that attacks the grains of the maize plant. In experiments carried out in Florida, untreated fields produced an average of only 1·6 worm-free ears of corn per 100; whereas those fields treated with carbaryl produced unblemished cobs at the rate of 86 per 100. Another crop plant, alfalfa (or lucerne as it is known in some parts of the world) is attacked by the alfalfa weevil. Applications of malathion and methoxychlor to test fields in Maryland some eighteen days before cutting increased the yield of the first and second cuttings from 1·6 to 2 tons per acre and the protein content by up to 20 per cent. An investment of $5 per acre yielded an increase of $20 in the value of the hay.

Cotton is a crop that is attacked by a wide range of insect pests. In Waco, Texas, plots treated with DDT yielded an average of 34 per cent more cotton over a six year period than did untreated plots. In Louisiana the increase was 41 per cent, in South Carolina 54 per cent. It is not surprising that the cotton-growing Mississippi Delta region has the highest rate of pesticide application in the world.

Harvest does not of course mark the end of the trouble. In cereal storage, many insects—mainly beetles—destroy 50 per cent of the world's production of food and degrade the rest. The struggle against pests is a very real one and it can only be waged with the help of pesticides, without which the losses would be enormous.

Insects such as aphids and leaf hoppers are also responsible for transmitting many plant diseases, particularly those of the virus type, and thus the control of insects can play a vital role in this field. It is not the complete answer. Many plant diseases, particularly those caused by fungi, are spread in other ways; so the fungus must be attacked directly. By controlling black rot in grapes, ferbam—a dithiocarbamate fungicide—has increased the yield per acre from 1,000 pounds to 8,000 pounds. Used against leaf curl in peaches, it raised the yield from one ton to seven tons. In the weed control field too there are some striking successes. In a Nebraska experiment, the use of 2,4-D raised the yield of forage plants by 254 per cent. In Maryland use of the herbicide chlorpropham on alfalfa fields increased the production from 3,000 pounds to 4,000 pounds per acre. One of the most spectacular yields was obtained by the use of dalapon on fields of bird's-foot trefoil: an increase of 4,825 per cent, from 80 to 3,860 pounds per acre. Chemical weed control has boosted the productivity of wheat, oats, barley and grazing land by 10–20 per cent. Herbicides have also been responsible for lowering the cost of weeding, as in strawberries where the cost is nine times less than hand weeding, and in the control of aquatic weeds where the cost is ten times less per mile than other methods. Pesticides have also played an important part in preserving the quality of produce. Sodium 2-phenylphenate is used on Florida oranges packed in polythene bags for shipping, and has reduced the proportion of oranges affected by decay from 25 to 7 per cent. Biphenyl is used on lemons, oranges and grapefruit in storage and this has effectively reduced decay from 10 to 5 per cent: the annual saving is calculated to amount to millions of dollars. All Californian grapes shipped are fumigated with sulphur dioxide. If they were not, the average rate of decay in a ten-week storage period would rise from 4 to 36 per cent, which would result in a loss of 192,000 tons of the annual 600,000 tons shipped.

The Yield

In forestry too pesticides play an important part. Several hundred thousand forest trees are killed each year by various kinds of bark beetle, which bore into the bark and spread fungal disease. When the bark beetles are ignored the effects can be devastating. Great wind storms in Colorado in the early 1940s flattened thousands of spruce trees. Due to the demands of the war effort, manpower and money were both short and the fallen spruce were neglected as potential breeding grounds. As a result there was a population explosion of the spruce beetle, leading to the destruction of 500,000 acres of forest and five thousand million feet of spruce timber before the epidemic was brought under control by the use of insecticides.

Pests can also attack farm animals; and anyone who has seen the misery that insects can cause to a cow might stop to consider that insecticide usage should be advocated from a humane point of view. In other words the life of a cow is just as important, from a humane point of view, as the life of a wild bird. The horn-fly —a bloodsucker—is about half the size of a housefly. It forms clusters round the back and shoulders of the cow, leaving it only to lay eggs in the cow-dung. A new generation is produced every two weeks and a single cow may be infested by as many as 4,000 flies. The only possible control is by the use of insecticides such as toxaphene, methoxychlor and malathion in the form of sprays and dips. Experiments carried out in Illinois showed that treated cattle showed a net gain in weight of 15 pounds per month. Milk production was also increased, and the butterfat content of the milk showed an increase of around 30 per cent.

Health gains

The World Health Organization states in its 1971 report that some 15–20 per cent of the world's production of 200,000–250,000 metric tons of DDT is currently used for the control of vectors of human disease, particularly malaria against which some 40,000 tons are used.

The report states that the advent of DDT has been the most important single factor in eradicating malaria from large parts of the world. Between 1959 and 1970, over 1,000 million people have been freed from the risk of malaria, which is consequently no longer a problem in the temperate parts of the world.

The author of this book recalls landing in Greece with a Mobile Malaria Unit in 1944–5 and finding malaria widespread throughout the country. Returning for a lecture tour in 1972 he found that the disease had been completely eradicated—due to the extensive use of DDT.

In Italy (whence came the name *mal aria*—bad air) there were more than 400,000 cases in 1945. The spraying of homes had been begun in 1944, together with other measures. Owing to these measures, the Pontine Marshes—a deadly malarial region for thousands of years—became the home of 100,000 healthy people. In Italy as a whole, malarial deaths have been practically unknown since 1948.

The effect on the health of the population is striking when one examines the malaria mortality figures reported from countries where malaria has been eliminated or where active campaigns are in progress. Following a massive DDT campaign in Sri Lanka the number of malaria cases dropped from about 3 million cases in 1946 to 110 in 1961 and the number of deaths dropped from 12,587 to zero over the same period. In Mauritius the number of deaths from malaria was 1,589 in 1948; by 1955 the number of deaths had dropped to three. In Venezuela the number of malarial patients fell from 800 thousand in 1943 to 800 in 1958.

The figures given for the fall in mortality rates are impressive, but as well as saving lives malaria eradication has also yielded enormous social gains and economic benefits. The improvement in health has broken the vicious circle of poverty and disease. It has contributed to the increased production of rice in many countries by increasing the work output of the labour force; and vast areas of land, formerly unused because of endemic disease, can be opened up for agricultural production now that the labour force can move in. Today over 1,000 million people are living in areas freed from endemic malaria and a further 329 millions are protected by DDT—an achievement unparalleled in the annals of public health.

Recent experiences in some countries should serve as an awful warning what could happen if DDT is withdrawn. Sri Lanka, where as we have noted malaria was reduced to hundreds of cases by 1961, now faces an epidemic of malaria with a total of over 2·5 million cases (close to the 1946 figures) already reported during the two years 1968 and 1969, following the stoppage of house spraying with DDT. The WHO considers that the withdrawal of DDT would provoke a major catastrophe in the annals of human health.

The Yield

Another example of what can happen when no control measures are used comes from Houston, Texas: this time involving the disease encephalitis—a disease which attacks the central nervous system and the brain. The virus of encephalitis is carried by the mosquito, which the area had no programme to control. Then migrating birds brought in the virus, soon the number of cases mounted and an epidemic was brewing. Spray programmes were therefore undertaken and the epidemic was controlled; but it left behind a toll of 38 dead and 1,000 seriously afflicted.

The evaluation of other insecticidal compounds for malaria eradication is in constant progress. In May 1971, WHO reported that of 1,300 insecticides evaluated only five are comparable to DDT. Two of these (propoxur and 'Landrin') are carbamates and three are organophosphorus compounds (fenitrothion, malathion, and fenthion). Fenthion is not quite safe enough, malathion not effective enough, and all are much more expensive than DDT.

Other diseases transmitted by mosquitoes

Yellow fever, dengue and other virus diseases carried by mosquitoes remain a serious challenge today. So too is Bancroftian filariasis, a disease spreading with great rapidity in many tropical countries. Its transmission is associated with poor sanitation, since the mosquito carrying the disease, *Culex pipiens fatigans*, breeds in highly polluted waters. It is resistant to control by chlorinated hydrocarbons, and this has until recently prevented the mounting of a large-scale control programme. The larvae may be controlled by certain organophosphorus compounds (notably fenthion and chlorpyrifos). National programmes are now being based on these chemicals in Asia; but suitable substitutes are required for use when resistance develops. To ensure interruption of the transmission of filariasis, a minimum of seven years uninterrupted vector control is necessary. The control of the urban vector of yellow fever and haemorrhagic dengue rests almost entirely upon the use of larvicides, of which difenphos is the most effective and safe. However, no satisfactory substitute is available at present.

Professor J.D. Gillet of Brunel University tells an interesting story of the interrelationships between yellow fever, monkey and man. In the forests of Central and South America a monkey-biting mosquito (*Haemagogus spegazzinii*) is responsible for the spreading of yellow fever among monkeys. The infected monkeys

156

usually die of the disease, and so do not have a direct link with man. But man cultivates the forests within which the mosquito thrives; the woodcutters are therefore bitten. If some mosquitoes are carrying the yellow fever virus then the woodcutters become infected—indeed, this is an occupational disease of woodcutters in South America. If the infected woodcutters return to town still incubating the disease they may, in time, be bitten by the local mosquito—*Aedes aegypti*—which will spread the disease through the community and so set an epidemic in motion.

There is an obvious need for continual vigilance in the control of this mosquito, even in areas free of yellow fever. It incidentally is of interest to note that in Africa monkeys infected with yellow fever remain active and healthy, and when they leave the forest to raid nearby banana trees, they may be bitten by the mosquito *Aedes simpsoni*—which can then bite and infect man and set up a focus of infection which may be picked up by *Aedes aegypti* and spread. In some areas of Latin America, the use of difenphos for *Aedes* eradication is supplemented by the application of a residual organophosphorus or carbamate insecticide in the vicinity of larval habitats. Applications of malathion have proved very effective in the control of adult *Aedes simpsoni* in Ethiopia, and for reducing the densities of adult *Aedes aegypti* in urban areas.

Chapter 12

PESTICIDES
AND THE ENVIRONMENT

As we have seen in the previous chapter, great benefits have accrued to mankind through the use of pesticides. Millions of people are now living who would formerly have died; many more millions are now living full and useful lives, whereas before the pesticide era they would have been crippled and incapacitated by disease; and agricultural production has been enormously increased. Whilst accepting the benefits, however, many people have expressed concern at the harm that pesticides might be doing to the environment—poisoning the soil, destroying wild-life and even presenting an unacceptable hazard to man himself.

The pesticide that has come under greatest attack is DDT, a very persistent chemical. Although it can be broken down by soil micro-organisms, the process is a very slow one and once in the soil DDT may persist for many years. If any area of the world ought to be free of pesticides, it would seem that the Antarctic should be such a place since no pests and relatively few animals or plants live on the continent and there is no knowledge of any application of pesticides there. But investigations sponsored by the National Science Foundation found 0·44 parts per million of DDT in the fish *Rhigophila*, 0·18 ppm in Adelie penguins, 0·15 ppm in Weddell seals and 2·8 ppm of DDE (a close chemical relative of DDT) in skuas (an Antarctic bird). Pesticides do therefore occur in the Antarctic, thousands of miles from the nearest point of use. How the pesticide got there is not clear. Some believe that the presence of DDT indicates that it is world-wide in its distribution; others that it may have come from Antarctic research stations, and not necessarily from a world-wide dispersion of DDT.

The amounts of DDT found in the tissues of wild animals, fish, and birds and their eggs varies from country to country and from region to region within a country, depending on the amount and the regularity of usage of DDT.

The fear has been expressed that birds may be killed and their eggs rendered sterile through the use of DDT. In Britain eggs of many species have been found to contain DDT in varying amounts. Most of those reported range from less than 1·0 ppm to 4 ppm. Montague's harrier and the peregrine, two birds of prey, were exceptional in having 17 ppm and 12·5 ppm respectively. Much higher figures have been reported from the USA, e.g. four egg-yolks of wild pheasants living in ricefields where DDT was used had an average content of 568 ppm and the amounts in herring-gulls' eggs ranged from 654 to 1,359 ppm. One interesting point to note is that many of these eggs, even with very high concentrations, contained live embryos and would presumably have given rise to live young. The subsequent fate of those young would, however, be uncertain, since there is some evidence that young hatching from eggs containing DDT may be more susceptible to adverse conditions than are 'normal' young. Thus, if they grew up under 'stress' conditions of, say, food shortage they would be more likely to succumb than normal young.

There is difficulty in relating these various levels to the amounts taken in by mother birds, but it might be a fair assumption that the amounts taken in by the birds were much higher than those found in the eggs. If so, we are led to the conclusion that birds have a high tolerance for DDT. At Strathclyde University, experiments have been carried out involving the injection of known amounts of DDT into the yolks of fertile hens' eggs. The highest concentration to be put into solution was 500 ppm; and this had no adverse effect on the hatching rate of the eggs, or on the subsequent development of the chickens.

DDT has also been found in fish, e.g. in California striped bass from various localities had a range of from 2–124 ppm and averaged 64 ppm. Shellfish samples from eight states in America contained DDT + DDE (a breakdown product of DDT) ranging from less than 0·008 ppm to 0·9 ppm. Oysters, which filter great quantities of water through their systems, can also accumulate pesticides from very low concentrations in their surroundings. One experiment which may be quoted showed that when oysters were immersed for 40 days in water containing 0·1 parts per thousand million of DDT the oysters stored quantities about 70,000 times the concentration in the water.

Water plants can also accumulate considerable quantities: a 4-acre marsh in Ohio was treated with 0·2 pounds per acre of DDT. Within a week, algae living in the water were showing 245 ppm in their tissues. Among the most important water plants are the microscopic phyto plankton which live in the sea. They supply us with much of the oxygen that we and other animals breathe; they also provide the foodstuffs upon which most marine animals including fish ultimately depend. Anything interfering with the growth of these plankton is of crucial importance; and there is considerable evidence that DDT at very low concentrations does this. Many scientists are concerned that if sufficient quantities of organochlorine insecticides found their way into rivers, estuaries and coastal waters, they could damage fish either directly or indirectly through their food supply. Insofar as Britain, the USA, Canada and some other countries are concerned, the feeling is that they may have acted in time to prevent serious damage to fish by organochlorine insecticides; but there is a fear that there may be real problems elsewhere, especially in developing countries.

On the land, worms can accumulate DDT. In the United States more than 157 ppm have been found in worms from areas treated with DDT for Dutch Elm disease control. Earthworms from three Mississippi cotton fields contained 9–32 ppm of DDT, whereas the surrounding soil had only 0·5–3·0 ppm.

Among animals, big game from Idaho and Washington contained less than 0·1 ppm, where little or no DDT had been used. Three months after treating forest with one pound of DDT per acre, deer from Colorado and New Mexico contained 0·05 to 43 ppm DDT residues in their fat with an average of 7·6 ppm. Elk from the same area contained 0–29 ppm DDT in their fat with a median of 7 ppm.

Man, too, has DDT residues within him. In Britain the average amount of DDT in body fat in the 1960s was 2 ppm. The Advisory Committee on Poisonous Substances, reporting in 1964, stated that:

Whilst it is perhaps undesirable on aesthetic grounds that our body fat should contain traces of DDT, there is no scientific evidence that the DDT does any injury while it is in the fat. It neither disturbs the activity of the fat tissue itself, nor is it free to injure the sites that are sensitive to its action.

A word of caution was, however, added—'new information about the biological action of DDT may emerge at any time'. As far as the writer is aware, no such new evidence has arisen up to the time of writing this book. Americans from the USA over the same period had around 12 ppm which incidentally means, as someone has observed, that Americans are not suitable for eating—they have more DDT in their bodies than they allow in any consumable foodstuffs. The acute toxicity of DDT is about the same as aspirin; some is retained in the body usually in the form of DDE. A substantial fraction is also excreted and the larger the dose then the higher the excretion. DDT, being very persistent, passes through food chains and in some cases actually tends to increase as a result.

But besides building up, DDT, though persistent, also slowly breaks down. This is something that is all too readily forgotten; for example, a number of cases are quoted by Dustman & Stickel in their article 'Pesticide Residues in the Ecosystem'. In a farm pond treated with DDT at the rate of 0·02 ppm, residues in mud declined to pre-treatment levels in eight weeks. However, even fifteen months after treatment residues in trout were still essentially the same as they were soon after treatment. On the other hand, residues in fish in the Connecticut river declined significantly from the autumn of 1963 to the spring of 1964. Elk in New Mexico sampled for DDT residues in their fat three months after DDT treatment of their area (at the rate of one pound per acre) contained 0–22 ppm. When sampled a year later, the elk showed a decline in DDT content of 81 per cent. It will be clear from all these examples that DDT is widely distributed through the plant and animal kingdoms, in soil and in water. The question is, does it do any harm? In this context, it is of relevance to quote a passage by Robert White-Stevens as a foreword to Volume 1 of *Pesticides in the Environment*:

Over the past 25 years when pesticide application to crops, livestock, industrial plants, homes, forests, swamps and indeed the total environment have increased over 1,000 fold there has not been one single medically annotated cause of sickness, let alone cancer or death, from the use of any (US) registered pesticide when it has been applied strictly in accordance with the label recommendations. There have been unfortunate accidents and deaths due to exposure of workers, children drinking, careless use etc. These are not indictments of pesticides but the irresponsibility of the user. DDT has a record which is practically unblemished so far as human life and human safety is concerned. Over the past 20 years some 1,000 million humans have been exposed to DDT to varying degrees. This aggregates approximately twenty thousand million man years of exposure without a single fatality or sickness, acute or chronic, with no report of cancer or death attributable to the pesticide by reliable medical reports.

It is also of considerable interest to quote the World Health Organization's budget presentation for the year 1971, announcing further use of DDT (40,000 tons annually) against malaria vectors:

No toxic effects have been observed during the last twenty years among the 200,000 persons who have been charged with the application of DDT within the malaria control campaign, nor among the 600–1,000 millions of people living in the abodes repeatedly treated with this insecticide. The only cases of poisoning with DDT have been persons who accidentally have ingested the insecticide but none of them has died.

It may be that President Nixon in the USA and Premier Trudeau in Canada were right to ban the use of DDT in their countries because of its persistence; but there is certainly no unequivocal evidence of toxicity even at high levels. In their report in 1964, the British Advisory Committee on Poisonous Substances used in Agriculture and Food Storage concluded that no restrictions should be placed on the current uses of DDT in agriculture, horticulture, home gardens and food storage practice, but that its use should be reviewed at the end of three years. This further review was published in 1969, and as a result the use of DDT has been very much reduced.

There are some who will say that we ought to withdraw DDT completely. It is not indispensable; indeed, it can be convincingly argued that any substance is dispensable—but some are more dispensable than others. We could withdraw penicillin because some people are allergic to it, but this would mean that we would condemn millions to death through disease. We could withdraw DDT because it is persistent, provided we are willing to condemn many millions of our fellow humans to death through malaria, typhus and the like because the alternatives that we have to offer are not as efficient as DDT and are much more expensive —so expensive that the poorer countries (where they are most needed) would not be able to afford them.

Regarding agriculture, a number of alternatives to the organochlorines are possible. In the United Kingdom voluntary curtailments of organochlorine usage followed governmental recommendations arising from the review of the pesticide situation in 1969, so that few approved uses now remain for DDT, aldrin and dieldrin. Heptachlor is not available in this country; nor is DDT for home or garden use. The search has intensified for alternative pesticides. Robert Gair of the Agricultural Development and Advisory Service had this to say at a conference in 1971:

I began by listing certain undesirable features of the persistent organochlorines which led to their severe curtailment in the United Kingdom (they persist in the environment, have deleterious effects upon beneficial and general wildlife, accumulate within the body tissues of predators and higher vertebrates and induce resistance problems in some of the pests which they formerly controlled). Let me end by recalling the virtues of DDT, aldrin, dieldrin and other persistent organochlorines. They were cheap, easy and safe to apply to crops and often gave wellnigh perfect control of many pests. By withdrawing so many uses for organochlorine insecticides, we move into an era when pest control in this country may be less effective for most crops, more hazardous to those who have to apply the new chemicals, and certainly more expensive. If it is not possible to educate the consumer public into accepting produce with a small amount of pest damage, then we may have to forsake reliance on chemical control methods alone and integrate the new non-persistent pesticides with non-chemical methods to give a degree of pest control to which the British farmer and public have become accustomed. Little progress has so far been made in developing integrated control methods in this country.

With regard to the banning or partial banning of DDT, Dr. R. Stevens complains that it is not possible to prove that DDT or indeed any chemical does no harm. Referring particularly to the American scene he states:

> The important point remains that if any useful compound, machine, tool, book or object in our society can be effectively destroyed, banned or disposed of upon unscientific, unjustifiable and largely fabricated and exaggerated evidence that is promoted by a vociferous, self-interested, minority then nothing can be regarded as safe, stable or established. It is recognised in all scientific research that the proof of the negative or null hypothesis is impossible namely that there is no difference between two or among more than two treatments whatever they may be. If there are no significant differences shown still the null hypothesis cannot be accepted. The public assume that science can prove the absence of a deleterious quality of a chemical as readily as it can prove its beneficial properties. They demand proof that a new compound is non-carcinogenic, non-teratogenic, non-mutagenic before its release. Such a demand is a complete impossibility to meet for even if each compound were studied on 1,000 species of living creatures each for 1,000 years the failure to observe these undesirable features does not constitute absolute proof of their total absence. The likelihood can however be computed by carefully controlled experiments—dose rates, routes of exposure and at differential frequencies and period durations.

Aldrin, dieldrin and heptachlor

In the early spring of 1956, large numbers of dead birds—particularly pigeons and pheasants—were found in various parts of England, especially in cereal-growing areas, at or soon after the sowing period. Sometimes the birds were observed to drop dead while in flight, and sometimes death was preceded by convulsions. Such incidents recurred in the following years, although there was some variability in the severity of attack. Then in 1960 there was circumstantial evidence that deaths had occurred amongst foxes, badgers and farm dogs and cats from the consumption of poisoned birds. Dieldrin, like DDT, is a fairly stable chemical, so birds and mammals feeding on birds poisoned by dieldrin were themselves being poisoned.

A series of investigations was undertaken on behalf of the Ministry of Agriculture, Fisheries and Food into the toxicity of four commonly-used seed dressings: dieldrin, aldrin, heptachlor and gamma-BHC. This was done by feeding pigeons known lethal quantities of these chemicals. The time of death ranged from $2\frac{1}{4}$ hours to 17 days, generally with convulsions. The birds were then analyzed and the amounts of pesticide noted. Bodies of birds were then collected from fields and similarly analyzed. The results indicated that the seed dressings aldrin, dieldrin and heptachlor were primarily responsible for the death of wild birds. A single pigeon can dig up great quantities of seed corn in a day, and it will probably receive a lethal dose of dieldrin in so doing.

Action followed soon afterwards. An announcement was made by the Minister of Agriculture, Fisheries and Food in the House of Commons that as from 1st January 1962, seed dressings containing dieldrin, aldrin and heptachlor would not be used for spring sowings but would only be used for dressing autumn and winter wheat where there was a real danger of attack from wheat bulb fly. The reason for the ban on dressing of spring sown seeds is that this is a time when wild seeds etc. are in very short supply and birds therefore eat a large amount of sown seed. In autumn there is an abundance of food available, so birds are much less likely to feed on sown seed.

A voluntary scheme was adopted by insecticide manufacturers, farmers and government advisers, which has proved to be successful. No mass death of birds has since occurred in successive springs—although the level of deaths among birds, especially in Scotland still proved to be unacceptably high and the British Ministry of Agriculture has recommended that aldrin and dieldrin should be withdrawn as autumn seed dressings from the whole of Britain from 1974 onwards. Gamma-BHC, which was also used as a seed dressing, was excluded from the ban because birds find this chemical to be quite distasteful and even when starving they are reluctant to eat it. When fed to birds in a 'disguised' form it invariably made them sick and they rejected it. Residues of gamma-BHC have been found in birds both in Britain and in America, but the amounts are so small as to render it harmless—though recent work suggests that there may be a marked reduction *post mortem*.

Apart from direct killing, there have been claims that small amounts of the organochlorine insecticides can seriously affect

165

the behaviour of birds so that they do not breed or incubate properly. This is of course an extremely difficult thing to prove. The presence of broken eggs in nests has been cited in support of this but another explanation has also been forthcoming. It has been shown that the thickness of the egg-shells of certain birds of prey—peregrine falcon, sparrowhawk and golden eagle—decreased significantly in 1946–50. This has been attributed to the use of organochlorine insecticides and the evidence that they interfere with the calcium metabolism of the mother bird is strong. The reduction in numbers of these birds of prey since 1950 is a fact and a cause for some concern. The sparrowhawk, for example, was until recently a common breeding bird throughout Britain. Today it is a rare bird in many areas. The reduction in numbers has been attributed to its being at the end of food chains and thus accumulating pesticides from the birds and other animals (e.g. sheep) on which it feeds. So far as birds of prey are concerned, the evidence is very strong that its decline was due to aldrin and dieldrin used as sheep dips; this has been confirmed by the recovery of populations following restrictions on the use of these substances.

The organophosphorus insecticides, which it will be recalled are related to nerve gases, include a number which are very poisonous to man. Parathion is very toxic, both to wildlife and to man, and a number of deaths have occurred due to the operators not using protective clothing. TEPP is also very toxic: one ounce could kill nearly 500 men and indeed any bird or mammal sprayed with parathion or TEPP would have very little chance of survival.

And yet there have been very few human or wildlife deaths due to the use of these chemicals. This is because they are hydrolyzed very quickly in the soil and converted to non-toxic substances. Most of the newer organophosphorus compounds are much less toxic and are also systemic, which means that there is less chance of anything other than the pest being affected. Malathion, which was introduced in 1950, and menazon in 1960, are only about one two-thousandth part as poisonous as TEPP against warm-blooded vertebrates.

Even with some systemics, however, great care has to be taken; demeton-S-methyl is poisonous to man and warm-blooded animals and spraying is usually carried out in situations where there is little chance of wildlife being present. Some organophosphorus compounds are also replacing dieldrin as a sheep dip because of their rapid breakdown.

It will be recalled that in Chapter 4, the term 'synergism' was used: the toxic effect of a combination of pesticides may be much greater than one would expect from purely additive effects. Unfortunately, synergism may also be effective in increasing the toxicity of pesticides against wildlife and man. It is now known that people exposed to apparently harmless doses of parathion may die on subsequent exposure to malathion (which is not normally a toxic substance, as it is readily broken down in the body). This is apparently brought about by traces of parathion interfering with the body-mechanism which brings about the breakdown of malathion (see p. 51). This means that malathion accumulates in the body and brings about death.

In work carried out on egg injection at Strathclyde University, it was found that organophosphorus compounds were far more toxic than organochlorine compounds. Not only did they cause a marked reduction in hatching rate, but if given in high enough concentrations they also caused a variety of deformities such as stunting of the body, shortening of the beak, entire bones missing in the legs and a shortening of the spine. But it should be noted that there are no reports of organophosphorus-contaminated eggs being found in the wild or of their contamination being induced by feeding the mother bird. This is due to the fact that these insecticides are not acceptable to birds; and, if they do happen to be eaten, they are rapidly broken down if in small quantities. If they do happen to be ingested in high quantities then the bird is killed and no eggs are laid.

Herbicides
DNOC is a very poisonous substance both when taken in through the mouth and through the skin. There have indeed been one or two deaths caused by protective clothing not being worn, and no doubt some small mammals and birds have died through being inadvertently sprayed in the cornfield. It is however rapidly broken down in the soil, so there is no food chain effect and as a herbicide it has largely been replaced by 2,4-D and MCPA. These are among the least poisonous of any substances known and there have been no reports of human deaths. The use of 2,4,5-T for defoliating forests in Vietnam did, it is claimed, lead to deformities and deaths; but this was shown to be due to an impurity, a

dioxin, in the 2,4,5-T. Birds and mammals covered with a spray of MCPA or 2,4-D have shown no ill-effects, and both these chemicals are fairly quickly broken down in the soil.

There is however an indirect effect on wildlife. Some insects feed almost exclusively on certain plants, and if those plants are destroyed by weedkillers then the source of food goes too. For example, five species of butterflies—the Peacock, the Red Admiral, the Map butterfly, the Comma butterfly and the Small Tortoiseshell—have the nettle as their food source. A plea is therefore made to farmers not to destroy all their nettles for the sake of tidiness. A few clumps left growing will provide for these beautiful butterflies. We should certainly all be the poorer should they go. Similarly, the Painted Lady feeds exclusively on thistles, the Glanville Fritillary on certain species of plantain, the Heath Fritillary on cow-wheat and the Marsh Fritillary on devil's-bit scabious. Many weeds are equally valuable sources of nectar for bees, e.g. dandelion, charlock, poppy and rosebay willowherb while plants such as coltsfoot, poppy and others are visited by bees for their pollen.

The monuron and simazine groups of herbicides are very persistent in the soil, but there is no record of their entering into food chains nor is there any record of their being toxic to animals or man. Paraquat is a very interesting herbicide in that it is very quickly inactivated in the soil—not by breakdown, as in the organophosphorus insecticides, but by being adsorbed on to soil colloids and held there until later broken down by soil micro-organisms. Paraquat is a very efficient and valuable herbicide but it is poisonous and care should be taken in its use not to expose the skin to it. In the Strathclyde egg injection tests, paraquat was found to be the most toxic of all the pesticides tested. Doses as low as 0·3 ppm in the egg killed the embryo. Subsequent tests have, however, shown that in order to get this amount into the egg, hens would have to be fed four times the concentration at present used for field spraying. There have recently been a number of deaths through the drinking of paraquat in mistake for mineral water. It causes a progressive pneumonia for which there is at yet no cure: the greatest care should therefore be taken in the storage of this chemical. In particular, beer or mineral water bottles should not be used for this purpose.

Fungicides

Most of the fungicides in use are of low mammalian toxicity, which pose no problem for wildlife and man. The two exceptions are DNOC, which we have already dealt with under herbicides; and the organo-mercury compounds, many of which are used as seed dressings. Here in Britain we do not seem as yet to have any problem with organo-mercurials. Poultry and game birds fed on dressed grain have shown no ill effects, though the practice is not one that should be encouraged.

At a Conference in November 1971, Professor Granhall reported that the organo-mercury fungicides have been under active investigation in Sweden. High mercury contents in seed-eating birds and birds of prey caused major alarm. Mercury compounds have been used for nearly forty years on seeds in autumn- and spring-sown cereals, sugar beet, other root crops, and many other vegetables both in liquid and dust forms. In 1958 the Institute of Veterinary Research published results of chemical analyses of wild birds. From the high mercury content and from feeding experiments with dressed grain, it was concluded that mercury treatment of seed could be held responsible for the high mercury content in grain-eating birds and their predators.

It was recommended (and later made law) that the dosage of mercury for spring-sown cereals should be reduced by a half, and that mercury treatment should be carried out only when seed analysis had revealed a significant infection of seed-borne disease. No restrictions have been imposed for winter cereals. In 1966, persistent mercury dressings were withdrawn and less stable mercurials were substituted and allowed to be used at full dosage rates. These regulations have produced remarkable changes in the mercury content of doves. In 1964, 70 per cent of doves shot had 1 ppm mercury in their livers and 30 per cent had 5 ppm; by 1967 only 8·5 per cent of analyzed doves had more than 1 ppm mercury in their livers and none over 5 ppm. Similar decreases were noted in dove-hawks, 30 per cent of which had over 10 ppm in 1963–4, but none over 3 ppm in 1967.

Poisons used against vertebrate pests

Although we have noted in Chapter 8 that some highly poisonous substances are used to control vertebrates, such as hydrogen cyanide for rabbits, strychnine for rats and moles and phosphine for moles, it would seem that they do little, if any, harm to any animals other than the target species. This is because the localized environment in which these pests operate makes it possible to direct the poison specifically against them. There is of course a risk to the operator, but this is taken care of by precautions in the wearing of protective clothing and masks where necessary. We have also noted that the dosages of rotenone used against rats are harmless to mammals and birds, though it is quite deadly to fish. It breaks down rapidly.

Pesticides and the soil

As mentioned earlier, great concern has been expressed in certain quarters that some pesticides may be 'poisoning the soil'. By this is meant that they may be killing off the micro-organisms on which the fertility of the soil depends or rendering the soil unfit for further crop growth. The soil is a mass of micro-organisms of all kinds—bacteria, fungi, algae and minute animals such as protozoa and nematodes. The number of bacteria in one gram of soil ranges from one million to several billion. It may help to put this into perspective a little by noting that 2 billion bacterial cells per gram of soil account for 0·2 per cent of the soil weight. This amounts to 4,000 pounds live weight of bacteria per acre of soil, to a depth of six inches. For comparison it may be noted that there are about one billion bacteria per gram of soured milk.

Micro-organisms play a most important part in the economy of the soil. Many of them break down plant and animal bodies, thus releasing nutrients that are available for uptake by plants. Without the micro-organisms, such bodies would accumulate and the earth would become choked in a very short space of time. Other bacteria utilize the nitrogen of the air—the term 'fixing' is sometimes used—and turn it into protein which will eventually become available for uptake by plants when it is broken down into its constituent ammonium and nitrate salts. Not all micro-organisms in the soil are beneficial—some are capable of causing plant diseases, but whether useful (as the vast majority are) or harmful, we cannot afford to ignore them.

There are three main routes by means of which pesticides reach the soil: (1) a number are sprayed on to the crop and a proportion of such chemicals inevitably reach the soil; (2) a number of pesticides are applied directly to the soil; and (3)

170

some pesticides are applied to seeds which are subsequently planted. This latter method is probably not of major importance from the point of view of the soil micro-organisms, since the application is fairly localized. The other two methods are, however, important to the agriculturist: (1), as mentioned above, the effect of pesticides on the soil micro-organisms; and (2) the part played by micro-organisms in decomposing pesticides in the soil and rendering them harmless.

It would appear to be well-established that the great majority of herbicides, when used at recommended rates, have little or no adverse effect on the micro-organisms of the soil. Some of them such as TCA may have an initial inhibitory effect, but this is followed by a rapid recovery of the soil population. Some herbicides such as dalapon and maleic hydrazide actually stimulate the growth of the soil micro-organisms, probably because they act as foodstuffs.

As regards fungicides and fumigants, these generally exert a greater lethal effect on the soil population than do other pesticides. This is as one would expect, since they are very often put into the soil to do just this. But after an initial reduction in numbers certain types, usually bacteria, return very quickly and reach numbers far in excess of those in untreated soil. In general the greater the initial kill, the greater the subsequent peak in numbers.

The results of various workers indicate that the effects from normal field rates of insecticides are not sufficiently great to cause a significant reduction in soil fertility. For example, only excessive applications of parathion are inhibitory; at normal rates it is rapidly decomposed by bacteria and the decomposition products stimulate the increase of various groups of bacteria including those capable of fixing nitrogen. Schradan—even at very high dose rates—has a stimulating rather than an adverse effect on the soil microbes, and there is a marked increase in the numbers of nitrogen-fixing bacteria.

Even repeated applications of insecticides have no adverse effect. Dr. Martin and his group in the United States applied five annual field applications of the following insecticides: aldrin, chlordane, DDT, dieldrin, endrin, heptachlor, gamma-BHC and toxaphene, at rates ranging from 5–20 pounds per acre, to two field soils. Tests showed that during the five year period and for about ten months after the fifth application, there was no effect on the numbers of soil bacteria and fungi, on the kinds of fungi, or on the ability of the soil population to break down organic matter.

There are various ways in which pesticides may be lost from the soil. These are: (1) by physical removal such as volatilization (turning from a solid into a gas), leaching (i.e. by being washed out of the soil), adsorption (i.e. by becoming bound on to soil colloids) and by being taken up unchanged by plants; and (2) by decomposition. This decomposition may be biological or non-biological; it may be brought about by microbes or it may be caused by some chemical agent such as acids occurring naturally in the soil. This, it will be recalled, is the means by which organophosphorus insecticides are degraded.

However, there is no doubt that the main breakdown of pesticides is biological. This can be determined by adding some poison such as sodium fluoride to the soil which will kill off all the microbes. If a pesticide is now added to the soil it will not be broken down in most cases, indicating that the presence of living microbes is necessary before this will take place. By isolating certain microbes from the soil and then growing them in the laboratory, it can also be demonstrated that these microbes will break down certain pesticides. A great number of such organisms have now been isolated. Most of them are specific, e.g. an organism that is capable of breaking down 2,4-D will not break down monuron. Another that will break down dalapon will not break down a carbamate, and so on.

Almost every pesticide seems to have one or more organism able to degrade it, and there are some soil microbiologists who believe that soil microbes are able to deal with almost any organic matter, including pesticides, that man cares to deposit in the soil. As has been said, they provide 'an inexhaustible waste disposal medium—a garbage can *par excellence*'. Somewhere in the soil there is a micro-organism, actual or potential, which can break down any organic compound.

Other soil microbiologists however take a different view: they maintain that this is not so, and that there are a considerable number of organic substances which, because of their peculiar molecular configuration, will not yield to biological breakdown processes. These substances include the outer surfaces of pollen grains and the various petroleum hydrocarbon deposits that have resisted breakdown in the soil for millions of years; and amber,

172

the fossil resinous material from insects which has withstood attack for perhaps 25 million years.

The point is perhaps a good one: microbes are not infallible. But we must not lose our sense of perspective. The examples given above are but a minute fraction of the organic substances in soil; the vast majority of pesticides are readily broken down. 2,4-D is decomposed in a matter of a few weeks, MCPA in a matter of months and dalapon in weeks or months depending upon the soil. Although there are a few pesticides that show considerable resistance, they are only a few. They include some of the chlorinated hydrocarbons such as chlordane, DDT and dieldrin which may persist for 3–15 years or longer in soil, although concentrations decrease with time; and the herbicides monuron and diuron, which can be found in the soil some twelve months after application.

However, as has been said, most pesticides are broken down readily in soil and the question arises where the microbes capable of attacking these 'foreign' pesticides come from. Are such organisms always present in the soil in small numbers, waiting only for the right pesticide to arrive so that they can use it and multiply? Or does the arrival of the pesticide induce in microbes already present the formation of enzymes capable of dealing with it? Some microbiologists hold one view, some the other. Only time will tell.

Summing up this particular section, one would say that pesticides in general do little if any harm to the soil microbes (in some cases they actually help them to grow), and although there are a few pesticides that are very persistent in soil the vast majority are readily broken down. There is little evidence that the soil is being poisoned. In addition, the effect of pesticides on the environment is constantly being monitored by government, industry, scientific bodies, universities and colleges, and by numerous voluntary bodies.

Government

Governmental activities in the pesticide field range from sponsoring research to legislating against the improper use of pesticides.

(a) Many ministries and departments are concerned with the use of pesticides, and research is carried out and sponsored by the Agricultural Research Council, the Medical Research Council, the Natural Environmental Research Council, and the Science

Research Council. These Councils have research institutes responsible to them, in which a great deal of work on pests and pesticides goes on. As an example, one might mention the A.R.C. Weed Research Organization in Oxford, where some 300 experiments are carried out annually on new and proposed herbicides. In addition the Councils make substantial grants to scientists in universities and elsewhere to pursue research work.

(b) The sale of the more dangerous chemicals, e.g. strychnine, is restricted by law; and farmers and growers are required to take certain safety precautions when using specified chemicals such as DNOC. Under the Pesticides Safety Precautions Scheme, manufacturers notify all new pesticides and new uses of existing pesticides before marketing them and provide detailed scientific evidence of their chemical and biological properties, so that hazards can be assessed by the Advisory Committee on Pesticides and other Toxic Chemicals. Recommendations for safe usage are issued under the scheme.

(c) The relevant government department is responsible for the Agricultural Chemicals Approval Scheme, which ensures that claims concerning approved products for the efficient control of pests, diseases or weeds match up to the instructions on the label.

(d) In 1960 the Nature Conservancy, the official body responsible for advising the government on conservation matters, set up a Toxic Chemicals and Wildlife Division with the intention of studying the effects of pesticides on wildlife. This is based on Monks Wood Experimental Station, Abbots Ripton, Huntingdon.

(e) Particular mention should also be made of the government-sponsored Infestation Control Laboratory which does such valuable work in the pest control field.

Industry

The amount of work done by industry to ensure safety of its products is not always appreciated by the general public. Thousands of highly promising pesticides have been rejected by industry on the grounds of toxicity to the environment and to man. A major concern in the development of any pesticide (see Chapter 3) is to ensure its safety for general release.

In addition, the bigger firms do an enormous amount of fundamental work on pesticides; and they also encourage such work in universities. It is all too often forgotten that it is certainly not in the interests of any company to release a toxic chemical: the repercussions are too great in financial and prestigious terms. Furthermore, pesticide manufacturers have to live in the same environment as the rest of us, and they too have children whose future is as important to them as ours are to us.

A major trade association, the British Pest Control Association, represents contractors and manufacturers. This is the body with which the government deals on matters concerning pesticide legislation and the operation of the Pesticide Safety Precautions Scheme.

Scientific bodies, universities and colleges
A number of scientific bodies like the Royal Society, the Royal Society of Edinburgh, the Association of Applied Biologists, the Society of Chemical Industry and many more sponsor meetings and symposia at which scientists can report on their scientific investigations. The British Crop Protection Council is particularly active in this respect, holding a major meeting at Brighton each year as well as involving itself with specialized topics. This body is a splendid example of voluntary co-operation between industry, government, advisers and research interests. It is a forum where all pesticide and wildlife interests are represented.

In many universities, polytechnics, technical colleges and agricultural colleges throughout the country, scientific work is in progress into various aspects of pesticide usage.

Voluntary bodies
Many voluntary bodies do excellent work into the effects of pesticides on the environment, both in restricted localities (for example, local natural history societies) and at a national level: among the most active of the latter are the British Trust for Ornithology, the Royal Society for the Protection of Birds and the Game Research Association.

It is the author's opinion that the British systems of approval, clearance and liaison among industry, government and nature conservancy interests are of a standard that many other countries might well envy.

Appendix I

FURTHER READING

CHAPTER ONE

1 PESTICIDES AND POLLUTION, K. Mellanby. New Naturalist Series, Collins, 1967.
2 PESTICIDES AND THE LIVING LANDSCAPE, R.L. Rudd. Univ. of Wisconsin Press, 1964.
3 THE INTERNATIONAL WILDLIFE ENCYCLOPAEDIA, 12 Vols, ed. M. & R. Burton. BPC Publishing Ltd., 1970.
4 READINGS IN ENTOMOLOGY, P. Barbosa & T.M. Peters. W.B. Saunders, 1972.
5 FUNDAMENTALS OF APPLIED ENTOMOLOGY, ed. R.E. Pfadt. Macmillan.
6 POMP AND PESTILENCE, Ronald Hare. Victor Gollancz, 1954.
7 INTRODUCTION TO PARASITOLOGY, A.W. Jones. Addison Wesley, 1967.
8 INSECTICIDE AND FUNGICIDE HANDBOOK, 4th Ed., ed. H. Martin. British Crop Protection Council, Blackwell Scientific Publications, 1972.
9 ANIMAL ECOLOGY IN TROPICAL AFRICA, D.F. Owen. Oliver & Boyd.
10 THE BIOLOGY OF WEEDS, ed. J. Harper. Blackwell Scientific Publications, 1959.
11 WEEDS AND ALIENS, Sir E. Salisbury. New Naturalist Series, Collins, 1964.
12 WEEDS OF THE WORLD, L.J. King. Leonard Hill Books, 1966.
13 *Significance of Weed Seeds in relation to Crop Production*, F.R. Horne *in* Proceedings of the British Weed Control Conference, 372–99, 1953.
14 WEED DISPERSAL AND PERSISTENCE, Sir E. Salisbury. Proc. Brit. Weed Control Conf., 289–300, 1954.
15 WEED CONTROL HANDBOOK, 5th Ed., Vol. I: Principles, ed. J.D. Fryer & S.A. Evans. Blackwell Scientific Publications, 1968.
16 MUSHROOMS AND TOADSTOOLS, John Ramsbottom. New Naturalist Series, Collins, 1953.

17 ALGAE AND FUNGI, C.J. Alexopoulos & H.C. Bold. Current Concepts in Biology, Macmillan, 1967.
18 PLANT PATHOLOGY, E.J. Butler & S.G. Jones. Macmillan, 1959.
19 THE ADVANCE OF THE FUNGI, E.C. Large. Jonathan Cape, 1940.
20 DISPERSAL OF FUNGI, C.T. Ingold. Clarendon Press, Oxford, 1953.

CHAPTER TWO

1 WEED DESTRUCTION, S.A. Evans, Blackwell Scientific Publications, 1962.
2 WEED CONTROL HANDBOOK, 5th Ed., 2 Vols., ed. J.D. Fryer & S.A. Evans, Blackwell Scientific Publications, 1968.
3 MANSON'S TROPICAL DISEASES, Philip Manson-Bahr, Cassell, 1972.
4 WEEDS OF THE WORLD, L.J. King. Leonard Hill, 1966.
5 PLANT PATHOLOGY, E.J. Butler & S.G. Jones. Macmillan, 1949.
6 READINGS IN ENTOMOLOGY, P. Barbosa & T.M. Peters. W.B. Saunders, 1972.

CHAPTER THREE

1 CHEMICALS FOR WEED CONTROL, G.S. Hartley & F.F. West. Pergamon Press, 1969.
2 *The Development of a New Pesticide*, R.A.E. Galley *in* SOME SAFETY ASPECTS OF PESTICIDES IN THE COUNTRYSIDE, ed. N.W. Moore & W.P. Evans. Joint ABMAC/Wild Life Education & Communications Committee.
3 *The Examination of New Agricultural Chemicals*, J.F. Newman *in* SOME SAFETY ASPECTS OF PESTICIDES IN THE COUNTRYSIDE.
4 *The Development of New Herbicides—an Industrial View*, R.A.E. Galley *in* Proc. 8th Brit. Weed Control Conf., Vol. 3, 770–9, 1966.
5 *A Viewpoint of the Weed Research Organisation*, J.D. Fryer *in* Proc. 8th Brit. Weed Control Conf., Vol. 3, 780–3, 1966.
6 *The Evaluation of a New Herbicide*, K. Carpenter, R.C. Amsden, K. Holly, J. Holroyd *in* WEED CONTROL HANDBOOK 5th Ed., ed. J.D. Fryer & S.A. Evans, 176–215. Blackwell Scientific Publications, 1968.
7 *Bioassay of Pesticide Residues*, Yun-Pei Sun *in* Advances in Pest Control Research 1, ed. R.L. Metcalf, 449–96. Interscience, N.Y., 1957.
8 *Chemical Analysis of Pesticide Residues*, M.S. Schechter & I. Hornstein *in* APCR 1, 1957.

Appendices

CHAPTER FOUR

1 CHEMICALS FOR PEST CONTROL, G.S. Hartley & T.F. West. Pergamon Press, 1969.
2 PESTICIDES IN THE ENVIRONMENT, Vol. I Part 1, ed. R. White-Stevens. Marcel Dekker Inc., N.Y., 1971.
3 *The Chemistry and Action of Organic Phosphorus Insecticides*, T.R. Fukuto *in* APCR 1, 1957.
4 PESTICIDES AND POLLUTION, K. Mellanby. New Naturalist Series, Collins, 1967.
5 *Joint Action in Insecticides*, P.S. Hewlett *in* APCR 3, 1960.
6 *The Status of Systemic Insecticides in Pest Control practice*, W.E. Ripper *in* APCR 1, 1957.
7 *Absorption and Translocation of Regulators and Compounds used to control Plant Diseases and Insects*, J.W. Mitchell, B.C. Smale, R.L. Metcalf *in* APCR 3, 1960.
8 *Mode of Action of Insecticides*, C.W. Kearns *in* Annual Review of Entomology, Vol. 1, 1956.
9 *On the Mode of Action of Insecticides*, F.P.W. Winteringham & S.E. Lewis *in* Ann. Rev. of Entomology, Vol. 4, 1959.
10 *Mode of Action of Insecticides*, C.C. Roan & T.L. Hopkins *in* Ann. Rev. of Entomology, Vol. 6, 1961.
11 *Mode of Action of Insecticides Exclusive of Organic Phosphorus Compounds*, P.A. Damk *in* Ann. Rev. of Entomology, Vol. 2, 1957.
12 *Chemistry and Mode of Action of Organophosphorus Insecticides*, E.J. Spencer & R.D. O'Brien *in* Ann. Rev. of Entomology, Vol. 2, 1957.
13 ORGANIC INSECTICIDES, R.L. Metcalf. 392, Interscience, 1955.
14 *Chemistry of Organic Insecticides*, T.R. Sukuto *in* Ann. Rev. of Emtomology, Vol. 6, 1961.
15 *Mode of Action of Carbamates*, J.E. Casida *in* Ann. Rev. of Entomology, Vol. 8, 1963.
16 *Wool Digestion and Mothproofing*, D.F. Waterhouse *in* APCR 2, 207–62, 1958.
17 *Joint Action in Insecticides*, P.S. Hewlett *in* APCR 3, 1960.
18 *Selective Toxicity of Insecticides*, R.D. O'Brien *in* APCR 4, 74–116, 1961.

CHAPTER FIVE

1 WEED CONTROL HANDBOOK, 5th Ed., 2 Vols., ed. J.D. Fryer & S.A. Evans. Blackwell Scientific Publications, 1968.
2 CHEMICALS FOR PEST CONTROL, G.S. Hartley & T.F. West. Pergamon Press, 1969.
3 PESTICIDES AND POLLUTION, K. Mellanby. Collins, 1967.

4 WEEDS OF THE WORLD—BIOLOGY AND CONTROL, L.J. King. Leonard Hill, 1966.
5 WEEDS AND ALIENS, Sir E. Salisbury. Collins, 1964.
6 THE CHEMISTRY AND MODE OF ACTION OF HERBICIDES, A.S. Crafts. Interscience, 1961.
7 WEED CONTROL AS A SCIENCE, G.C. Klingman. John Wiley & Sons, 1966.
8 THE PHYSIOLOGY AND BIOCHEMISTRY OF HERBICIDES, L.J. Audus. Academic Press, 1964.
9 *The Chemistry and Mode of Action of Herbicides,* A.S. Crafts *in* APCR 1, 39–80, 1957.
10 *Chemistry and Herbicidal Properties of Triazine Derivatives,* H. Gysin & E. Knüsli *in* APCR 3, 289–358, 1960.
11 *Chemical Warfare,* E. Long *in* Farmers Weekly, Jan. 21st 1972.
12 *The Control of Aquatic Weeds,* T.O. Robson *in* M.A.F.F. Bulletin, 194, 1973.

CHAPTER SIX

1 *Problems & Progress in the use of Systemic Fungicides,* E. Evans *in* Proceedings 6th British Insecticide and Fungicide Conference, 3, 1971.
2 CHEMICALS FOR PEST CONTROL, G.S. Hartley & T.F. West. Pergamon Press, 1969.
3 SCIENTIFIC PRINCIPLES OF CROP PROTECTION, 6th Ed., H. Martin. Edward Arnold, 1973.
4 *The Chemistry and Biology of Pesticides,* R.L. Metcalf *in* Pesticides in the Environment, Vol. 1 Pt. 1, 1971.
5 *Plant Pathology,* E.J. Butler & S.G. Jones. Macmillan, 1949.
6 *Symposium on Systemic Fungicides in* Pesticide Science Vol. 2 No. 5, Sept./Oct. 1971.
7 SYSTEMIC FUNGICIDES, ed. R.W. Marsh. Longman, 1972.

CHAPTER SEVEN

1 *The Application of Herbicides,* S.A. Evans *et al in* WEED CONTROL HANDBOOK 1, 101–51, 1968.
2 CHEMICALS FOR PEST CONTROL, G.S. Hartley & T.F. West. Pergamon Press, 1969.
3 SCIENTIFIC PRINCIPLES OF CROP PROTECTION, 6th Ed., H. Martin. Edward Arnold, 1973.
4 WEED CONTROL AS A SCIENCE, G.C. Klingman. John Wiley & Sons, 1966.

5 *Manual of Fumigation for Insect Control*, M.A.U. Munro *in* Agricultural Studies No. 56, FAO Rome, 1961.
6 *Some Parameters in the use of Fumigants*, B. Berck *in* World Review of Pest Control 3, 156, 1964.
7 *Spraying 500 acres a day*, T. Collier *in* Big Farm Management 4, 39–41, 1971.
8 *Waterless Spraying from the Air*, *in* Technical Monograph No. 2, CIBA Agrochemicals Division, December 1969.

CHAPTER EIGHT

1 THE RABBIT, H.V. Thompson & A.N. Worden. New Naturalist Special Vol. 13, Collins, 1956.
2 *Post-Myxomatosis Rabbit Populations in England and Wales*, H.G. Lloyd. EPPO Pub. Ser. A. No. 58, 197–215, 1970.
3 *The Control of Rodents*, E.W. Bentley *in* WHO Chronicle 21, No. 9, 363–8.
4 *Rats Resistant to Warfarin*, D. Drummond *in* New Scientist, 771–772, 23rd June, 1966.
5 *Warfarin Resistant Rats*, J.H. Greaves *in* Agriculture, 107–110, March 1970.
6 *Control of Wild Mink*, H.V. Thompson *in* Agriculture, 114–116, March 1967.
7 *Research on Grey Squirrel Control in Britain*, K.D. Taylor & H.G. Lloyd. EPPO Pub. Ser. A, No. 58, 185–96, 1970.
8 *A Campaign against Feral Coypus* (Myocastor coypus *Molinia*) *in Great Britain*, J.D. Norris *in* Journal of Applied Ecology 4, 191–9, May 1967.
9 *Wood Pigeons*, J.D. Norris *in* Agriculture, 587–91, December 1968.
10 THE WOOD PIGEON, R.K. Murton. New Naturalist Special Vol. 20, Collins.
11 *Improved Stupefying Baits for the Control of Town Pigeons*, R.J.P. Thearle *in* International Pest Control, March/April 1971.
12 *The Starling Menace*, T. Brough *in* The Grower, July 1967.
13 *The Dispersal of Starlings from Woodland Roosts and the Use of Bio-Acoustics*, T. Brough *in* Journal of Applied Ecology 6, 403–10, 1969.
14 THE HOUSE SPARROW, D. Sumners-Smith. New Naturalist Special Vol. 19, Collins, 1964.
15 *Farm and Granary Pests*, I. Thomas *in* Journal of the Royal Agricultural Society of England, Vol. 127, 187–205, 1966.
16 *Current developments in Pest Control*, *in* Symposium Caxton Hall, London, The Royal Society of Health, 18th February 1970.
17 CHEMICALS FOR PEST CONTROL, G.S. Hartley & T.F. West. Pergamon Press, 1969.

18 PESTICIDES AND POLLUTION, K. Mellanby. Collins, 1967.
19 THE SCIENTIFIC PRINCIPLES OF CROP PROTECTION, 6th Ed., H. Martin. Edward Arnold, 1973.
20 PESTICIDES IN THE ENVIRONMENT, ed. R. White-Stevens. Marcel Dekker, 1971.
21 THE INTERNATIONAL WILDLIFE ENCYCLOPAEDIA Vols. 1–20, eds. D.M. Burton & R. Burton. BPC Publishing Ltd., 1970.
22 THE CONTROL OF INJURIOUS ANIMALS, J.M. Cherrett, J.B. Ford, I.V. Herbert, A.J. Probert. English Universities Press, 1971.

CHAPTER NINE

1 *Biological Problems arising from the Control of Pests and Diseases, in* Symposia of the Institute of Biology No. 9, 1960.
2 WEEDS OF THE WORLD, L.J. King. Leonard Hill Books, 1966.
3 PEST CONTROL: BIOLOGICAL, PHYSICAL AND SELECTED CHEMICAL METHODS, eds. W.W. Kilgore & R.L. Doutt. Academic Press, 1967.
4 BIOLOGICAL CONTROL OF INSECTS, PESTS AND WEEDS, ed. P. de Bach. Chapman & Hall, 1964.
5 *The Status of Systemic Insecticides in Pest Control Practices*, W.E. Ripper *in* APCR 1, 1957.
6 *Biological Control of Pests & Weeds*, D.F. Waterhouse & F. Wilson *in* Science Journal, 31–7, December 1968.
7 *Predacious Fungi and the Control of Eelworms*, C.L. Duddington *in* Viewpoints in Biology I, eds. J.D. Carthy & C.L. Duddington, 151–200. Butterworth, 1962.
8 BENEFICIAL INSECTS, Bulletin 20. Ministry of Agriculture, Fisheries and Food, London.

CHAPTER TEN

1 PEST CONTROL: BIOLOGICAL, PHYSICAL AND SELECTED CHEMICAL METHODS, eds. W.W. Kilgore & R.L. Doutt. Academic Press, 1967.
2 *Insect Chemosterilants*, Alexej. B. Borkovec *in* APCR 7, 1966.
3 *Male Sterilization for the Control of Insects*, R.C. Bushland *in* APCR 3, 1960.
4 *Recent Developments in Chemical Attractants for Insects*, N. Green, M. Beroza & S.A. Hall *in* APCR 3, 1960.
5 *Insect Hormone Mimics and their use in Insect Control*, F.S. Downing, N. Punja and C.N.E. Ruscoe *in* Reports on the Progress of Applied Chemistry Vol. IV, 446–57. Society of Chemical Industry, 1970.
6 READINGS IN ENTOMOLOGY, P. Barbosa & T.M. Peters. W.B. Saunders, 1972.

N

Appendices

CHAPTER ELEVEN

1 FAO Symposium on Crop Losses. Rome, 2nd–6th October, 1967.
2 *Mankind and Civilization at Another Crossroads*, N.E. Borlaug. McDougall Memorial Lecture Conference, FAO, 8th November 1971.
3 *'That We may Live'*, J.L. Whitten. Van Nostrand, 1966.
4 *Symposium on Problems of Mosquito Control, in* Pesticide Science, Vol. 3, No. 4, August 1972.
5 MOSQUITOES, J.D. Gillett. Weidenfeld & Nicolson, 1971.
6 *The Place of DDT in Operations against Malaria and Other Vector-Borne Diseases, in* Official Records of WHO, No. 190, April 1971.
7 *Pesticides in Perspective: No. 4, Winter, in* CIBA-Geigy Journal, 1971/72.
8 PANS, Vol. 16, No. 2, June 1970.
9 PESTICIDES AND THE LIVING LANDSCAPE, R.L. Rudd. Faber & Faber, 1964.
10 *News & Pesticide Review*, Vol. 26, No. 4, April 1968.
11 *News & Pesticide Review*, Vol. 21, No. 6, August 1963.
12 *Skirmishes in the War on Want*, L. Broadbent. Paper based on a public lecture given at the University of Bath in November 1971.

CHAPTER TWELVE

1 *Pesticides in the Antarctic*, J.L. George & D.E.H. Frear, *in* Pesticides in the Environment and their Effects on Wildlife, Journal of Applied Ecology, Vol. 3 Supplement, 155–67, 1966.
2 *An Investigation of the Toxicity of Insecticides to Birds' Eggs using the Egg-Injection Technique*, J.F. Dunachie & W.W. Fletcher *in* Annals of Applied Biology, 64, 409–23.
3 Fifth Report of the Joint Committee of the British Trust for Ornithology and the Royal Society for the Protection of Birds on Toxic Chemicals, 1963–4.
4 *The Pesticide Game Survey*, J.S. Ash. Game Research Association Report for 1965.
5 *Pesticide Wildlife Investigations in California*, E.G. Hunt & J.O. Keith *in* Proceedings of the Second Annual Conference on the use of Agricultural Chemicals in California, 1962.
6 *'Reproduction in a Population of Herring Gulls* (Larus argentus) *contaminated by DDT'* in *Pesticides in the Environment and their Effects on Wildlife*, J.A. Keith, Journal of Applied Ecology, Vol. 3 Supplement, 1966.
7 *Pesticides and their Effects on Soils and Water*, ASA Special Publication No. 8. Soil Science Society of America, November 1966.
8 *Review of the Persistent Organochlorine Pesticides*, report by the

Advisory Committee on Poisonous Substances used in Agriculture and Food Storage. 196, HMSO, 1964.

9 *The Effects on Birds of certain Chlorinated Insecticides used as Seed Dressings*, E.E. Turtle and colleagues *in* Journal of the Science of Food and Agriculture, 8, 567–77, 1963.

10 PESTICIDES AND POLLUTION, K. Mellanby. Collins, 1967.

11 *Environmental Aspects of using Persistent Pesticides*, I. Granhall *in* Proc. 6th Brit. Insect. and Fung. Conf., 1971.

12 WEEDS AND ALIENS, Sir Edward Salisbury. Collins, 1964.

13 *Organochlorine Alternatives—a Review of the Present Position in the U.K.*, Robert Gair *in* Proc. 6th Brit. Insect. and Fung. Conf.

14 *Some Safety Aspects of Pesticides in the Countryside, in* Proceedings of a Conference at the British Museum (Natural History), London, 20th November 1967.

15 HERBICIDES AND THE SOIL, eds. E.K. Woodford and G.R. Sagar. Blackwell Scientific Publications, 1960.

16 SILENT SPRING, Rachel Carson. Penguin Books, 1966.

17 PESTICIDES IN THE ENVIRONMENT, ed. R. White-Stevens, Vol. 1, Pt. 1. Marcel Dekker Inc., 1971.

18 WHERE WE ARE AT—WITH DDT AND ALL THAT, Norman Moore. Development Forum, May 1973.

19 *A Synopsis of the Pesticide Problem*, Norman Moore *in* Advances in Ecological Research 4, Academic Press.

JOURNALS AND PERIODICALS

The following journals and periodicals include articles and papers concerned with scientific aspects of pest control, and they should prove useful to readers who may wish to pursue aspects of the subject.

1. *Insects and Related Organisms*
Annual Review of Entomology
Bulletin of Entomological Research
Canadian Entomologist
Canadian Insect Pest Review
Entomologia, Experimentalis et Applicata
Journal of Economic Entomology
Journal of Entomology Series A. General Entomology
Journal of Medical Entomology
Mosquito News
Proceedings of the British Insecticide and Fungicide Conferences
Review of Applied Entomology
Transactions of the Royal Entomological Society, London

Appendices

2. *Weeds*
Journal of Weed Science (formerly 'Weeds')
Proceedings of the British Weed Control Conferences
Reports of the Agricultural Research Council Weed Research
    Organisation
Technical Reports of the Agricultural Research Council Weed Research
    Organisation
Weed Abstracts
Weed Research

3. *Plant Diseases*
Annual Review of Phytopathology
Contributions of the Boyce Thompson Institute
Phytopathology
Plant Disease Reporter
Proceedings of the British Insecticide and Fungicide Conferences
Review of Plant Pathology

4. *General*
Advances in Pest Control Research
American Journal of Botany
Annual Review of Microbiology
Annual Review of Plant Physiology
Annals of Applied Biology
Annual Report Scottish Horticultural Research Station
Canadian Journal of Botany
Canadian Journal of Microbiology
Chemistry & Industry
Environmental Pollution
Horticultural Science
International Pest Control
Journal of Agricultural Research
Journal of Animal Ecology
Journal of Applied Microbiology
Journal of Experimental Botany
Journal of General Microbiology
Journal of Horticultural Research
Journal of the Science of Food & Agriculture
Pest Control
Pesticide Science
Physiologia Plantarum
Proceedings of the Society of Chemistry & Industry
Recent Advances in Pest Control
Reports of the National Vegetable Research Station

Reports of Rothamsted Experimental Station
Residue Reviews
Science
Soil Science
Technical Bulletins of the Ministry of Agriculture, Fish & Food
World Review of Pest Control

In addition to the above, many organizations (such as the European Weed Control Council and the Weed Science Society of America) hold meetings and symposia concerned with aspects of pest control, of which proceedings are published.

Appendix 2

COMMON AND SCIENTIFIC NAMES OF PESTS

(1) INSECTS AND RELATED ORGANISMS

Common Name	*Scientific Name*
Alfalfa weevil	*Hypera postica*
Ants	Formicoidea super family
Aphids	Aphididae family
Argentinian moth borer	*Cactoblastis cactorum*
Bark beetle	*Ips confusus*
Bee	*Apis mellifera*
Biting midges	*Culicoides*
Black scale (citrus)	*Saissetia oleae*
Black bean aphid	*Aphis fabae*
Blood fluke	*Schistosoma* sp.
Blowfly	*Calliphora vomitoria*
Body louse	*Pediculus humanus*
Boll weevil	*Anthonomus grandis*
Bollworm (cotton)	Caterpillar of *Heliothis armigera*
Buffalo gnat	*Simulium* spp.
Bugs	Hemiptera order
Butterflies, moths	Lepidoptera order
Cabbage aphid	*Brevicoryne brassicae*
Cabbage root fly	*Erioischia brassicae*
Capsids	Miridae family
Carabid beetle	Carabidae family
Carpet beetle	*Anthrenus scrophulariae*
Carrot fly	*Psila rosae*
Cattle tick	*Margaropus annulatus* and seven other *genera*

Common Name	*Scientific Name*
Chiggers	Mites of the genus *Trombicula*
Citrophilus mealybug	*Pseudococcus fragilis*
Clothes moth	Tineidae family
Cockroach	Dictyoptera order
Codling moth	*Cydia pomonella*
Colorado beetle	*Leptinotarsa decemlineata*
Comma butterfly	*Polygonia c-album*
Corn earworm	*Heliothis zea*
Corn leaf hopper	*Peregrinus maidis*
Cotton aphid	*Aphis gossypii*
Cottony cushion scale	*Icerya purchasi*
Cutworms	*Noctuid larvae*
Desert locust	*Schistocerca gregaria*
Dog ticks	*Dermacentor reticulatus, Rhipicephalus sanguineus* and *Ixodes holocyelus*
Dwarf pond snail	*Limnaea trunculata*
Earwig (common)	*Forficula auricularia*
Eelworm (plant)	Nematoda class, Tylenchida order
Elm bark beetle	*Scolytus scolytus* and *S. multistriatus*
European bug	*Pyrrhocopis apterus*
European pine sawfly	*Neodiprion sertifer*
Giant snail	*Achatina fulica*
Glanville fritillary	*Melitaea cinxia*
Gnat	Culicidae family
Grain weevil	*Sitophilus* spp.
Grape-leaf hopper	*Erythroneura variabilis*
Grape-leaf skeletonizer	*Harpisina brillians*
Grasshopper	Acridiidae family
Greenbottle fly	*Phaenicia sericata*
Gypsy moth	*Lymantria dispar*
Harvest mite	*Trombicula autumnalis*
Heath fritillary	*Mellicta athalia*
Hedgehog tick	*Ixodes hexagonus*
Hessian fly	*Mayetiola destructor*
Hop aphid	*Phorodon humuli*
Horn-fly	*Haematobia irritans*
Housefly	*Musca domestica*
Ichneumon fly	Ichneumonidae family
Intestinal fluke	*Fasciolopsis buski*
Itch mite	*Sarcoptes scabei*
Japanese beetle	*Popillia japonica*

Common Name	Scientific Name
Ladybird	*Novius cardinalis*
Leaf hoppers	Jassidae family
Leatherjackets (larvae of crane fly)	Tipulidae family
Lice	Anoplura order
Liver fluke	*Fasciola hepatica*
Louse (human)	*Pediculus humanus*
Lung fluke	*Paragonimus westermani*
Map butterfly	*Araschnia levana*
Marsh fritillary	*Euphydryas aurinia*
Mealworms	*Tenebrio* spp.
Mealybug	*Planococcus* spp.
Mediterranean fruit fly	*Ceratitis capitata*
Melon fly	*Dacus curcubitae*
Mexican fruit fly	*Anastrepha ludens*
Milkweed bug	*Oncopeltus fasciantus*
Millipedes	Diplopoda class
Mirid bug	*Cytorrhinus mundulus*
Mites	Acari order
Narcissus fly	*Merodon equestric, Eumerus tuberculatus* and *E. strigatus*
Oak eggar insect	*Lasiocampa quercus*
Olive fly	*Dacus oleae*
Oriental fruit fly	*Dacus dorsalis*
Painted Lady	*Vanessa cardui*
Pacific coast wireworm	*Limonius canus*
Peach-potato aphid	*Myzus persicae*
Peacock butterfly	*Inachis io*
Pine processionary caterpillar	*Thaumetopoea pityocampa*
Potato root eelworm	*Heterodera rostochiensis*
Praying mantid	*Mantids religiosa*
Predaceous ant	*Oecophylla smaragdina*
Rabbit flea	*Spilopsyllus cuniculi*
Raspberry beetle	*Byturus tomentosus*
Rat flea	*Xenopsylla cheopis*
Red Admiral	*Vanessa atalanta*
Red spider mite	Acarina order
Sawfly	Sub-group of Hymenoptera order
Scale insects	Coccidae family
Scorpion	Scorpionida order, e.g. *Chactas* sp.
Screw worm	*Callitroga hominivorax*
Sheep tick	*Ixodes ricinus*
Silkworm	*Bombyx mori*
Slug	*Limax* sp. and many others

Common Name	*Scientific Name*
Small Tortoiseshell	*Aglais urticae*
Snail	Mollusca order
Soft scale	*Coccus hesperidum*
Spiders	Araneae order
Stable fly	*Stomoxys calcitrans*
Stick insects	Phasmidae family
Stink-bug	Pentatomidae family
Strawberry aphid	*Pentatrichopus fragaefolii*
Sugar beet eelworm	*Heterodera schachtii*
Termites	Isoptera order
Thrips	Thysanoptera order
Ticks	Ixodoidea super family
Tsetse fly	*Glossina* spp.
Vedalia beetle	*Rodolia cardinalis*
Walnut aphid	*Chromaphis juglandicola*
Wasp	*Vespula vulgaris* and *V. germanica*
Wheat bulb fly	*Leptohylemyia coarctata*
Wheat stem sawfly	*Cephus pygmaeus*
Wireworm	Larvae of *Agriotes* and *Athous* spp.

(2) PLANTS

American Chestnut	*Castanea dentata*
Annual meadow grass	*Poa annua*
Autumn crocus	*Colchicum autumnale*
Autumnal hawkbit	*Leontodon autumnalis*
Barberry	*Berberis vulgaris*
Bent grass	*Agrostis* spp.
Bermuda grass	*Cynodon dactylon*
Bindweed (larger)	*Calystegia sepium*
Bindweed (black)	*Polygonum convolvulus*
Bird's foot trefoil	*Lotus* spp.
Bishopweed	*Aegopodium podograria*
Blackberry	*Rubus fruticosus* agg.
Blackgrass	*Alopecurus myosuroides*
Black nightshade	*Solanum nigrum*
Bracken	*Pteridium aquilinum*
Bramble	*Rubus fruticosus*
Broad-leaved dock	*Rumex obtusifolius*
Bulbous buttercup	*Ranunculus bulbosus*
Buttercup	*Ranunculus* spp.
Canadian water weed	*Elodea canadensis*

Common Name	Scientific Name
Chamomile	*Matricaria chamomilla*
Charlock	*Brassica sinapis*
Chickweed	*Stellaria media*
Cleavers	*Galium aparine*
Coltsfoot	*Tussilago farfara*
Corncockle	*Agrostemma gigatho*
Cornflower	*Centaurea cyanus*
Corn marigold	*Chryanthemum segetum*
Corn poppy	*Papaver rhoeas*
Couchgrass	*Agropyron repens*
Creeping buttercup	*Ranunculus repens*
Creeping thistle	*Cirsium arvense*
Crowsfoot	*Ranunculus* spp.
Curled-leaf dock	*Rumex crispus*
Daisy	*Bellis perennis*
Dandelion	*Taraxacum officinale*
Darnel	*Lolium temulentum*
Dock	*Rumex* spp.
Dodder	*Cuscuta* spp.
Elm	*Ulmus* spp.
European Chestnut	*Castanea sativa*
Eyebright	*Euphrasia officinalis*
Fat hen	*Chenopodium album*
Floating fern	*Salvinia auriculata*
French bean	*Phaseolus vulgaris*
Frogbit	*Hydrocharis morsus-ranae*
Fumitory	*Fumaria officinalis*
Garden pea	*Pisum sativum*
Gorse	*Ulex europaeus*
Groundsel	*Senecio vulgaris*
Hazel	*Corylus avellana*
Hemlock	*Conium maculatum*
Hogweed	*Heracleum sphondylium*
Holly	*Ilex aquifolium*
Horsetails	*Equisetum* spp.
Johnson grass	*Sorghum halepense*
Klamath weed	*Hypericum* sp.
Lesser celandine	*Ranunculus ficaria*
Lettuce	*Lactuca sativa*
Lucerne	*Medicago sativa*
Marsh thistle	*Cirsium palustre*
Mayweed	*Matricaria* spp.
Nettle	*Urtica* spp.

Common Name	*Scientific Name*
Oak	*Quercus* spp.
Plantains	*Plantago* spp.
Polygonum	*Polygonum* spp.
Pondweed	*Potamogeton* spp.
Prickly pear	*Opuntia* spp.
Primrose	*Primula vulgaris*
Quinine Tree	*Cinchona* spp.
Ragwort	*Senecio jacobaea*
Red clover	*Trifolium pratense*
Red fescue	*Festuca rubra*
Red rattle	*Pedicularis palustris*
Reed	*Phragmites communis*
Rhododendron	*Rhododendron* spp.
Rosebay willowherb	*Chamaenerion angustifolium*
Runch	*Brassica alba*
Rushes	*Juncus* spp.
Sainfoin	*Onobrychis viciifolia*
Sea leek (squill)	*Urginia maritima*
Sedge	*Carex* spp.
Sheep sorrel	*Rumex acetosella*
Shepherds purse	*Capsella bursa-pastoris*
Smooth-stalked meadow-grass	*Poa pratensis*
Sorrels	*Rumex acetosa; R. acetosella*
Sow thistle	*Sonchus* spp.
Spear thistle	*Cirsium vulgare*
Speedwell	*Veronica* spp.
Spurrey	*Spergula arvensis*
St. John's wort	*Hypericum perforatum*
Subterranean clover	*Trifolium subterraneum*
Sweet clover	*Melilotus alba*
Thistle	*Cirsium* spp.
Tumbleweed	*Salsola pestifer* and *Amaranthus graecizans*
Water dropwort	*Oenanthe crocata*
Water hyacinth	*Eichhornia crassipes*
Water milfoil	*Myriophyllum spicatum*
White clover	*Trifolium repens*
White mustard	*Sinapis alba*
Wild carrot	*Daucus carota*
Wild oats	*Avena fatua* and *A. ludoviciana*
Wild onion	*Allium vineale*
Yellow mustard	*Sinapis arvensis*
Yellow rattle	*Rhinanthus* spp.

Appendices

Common Name of Disease	Scientific Name of Causative Organism
Apple scab	*Venturia inaequalis*
Black rot of grapes	*Guignardia bidwellii*
Black rust	*Puccinia graminis*
Brown rot of fruits	*Sclerotinia fructigena*
Bunt of wheat	*Tilletia caries* and *T. foetida*
Citrus canker	*Xanthomonas citri*
Club-root	*Plasmodiophora brassicae*
Coffee-leaf disease	*Hemileia vastatrix*
Cucumber mildew	*Erysiphe cichoracearum*
Downy mildew of vine	*Plasmopara viticola*
Dry-rot fungus	*Merulius lacrymans*
Dutch Elm disease	*Ceratostomella ulmi*
Honey fungus	*Armillaria mellea*
Leaf-curl of peach	*Taphrina deformans*
Leaf-spot of celery	*Septoria apii* and *S. apii-graveolentis*
Leaf-spot of sugar beet	*Cercospora beticola*
Plague bacterium	*Pastuerella pestis*
Potato blight	*Phytophthora infestans*
Powdery mildews	*Erysiphales* order
Powdery mildew of vine	*Uncinula necator*
Ringspot of potatoes	*Bacterium sepedonicium*
Ringworm	*Tinea* spp.
Silver leaf of stone fruit	*Stereum purpureum*
Smuts	*Ustilago* spp.
Stinking smut of wheat	*Tilletia foetida*
Yeasts	*Saccharomycetes*

Common Name	Scientific Name
Adelie Penguin	*Pygoscelis adeliae*
Alligator	*Alligator missisipiensis*
Armadillo	Three *genera*: *Chaetophractus, Dasypus* and *Chlamyphorus*
Badger	*Meles meles* (European) and *Taxidea taxus* (American)

Common Name	*Scientific Name*
Black Rat	*Rattus rattus*
Brown Rat	*Rattus norvegicus*
Buzzard	*Butoe buteo*
Carp	*Cyprinus* sp.
Common Seal	*Phoca vitulina*
Coypu	*Myocastor coypus*
Deer (Red)	*Cervus elaphus*
Dove (Rock)	*Columba livia*
Dove-Hawk (Merlin)	*Falco columbarius*
Elk	*Cervus canadensis*
European Mole	*Talpa caeca*
Fox (Red)	*Vulpes vulpes*
Golden Eagle	*Aquila chrysaëtos*
Gouramy Fish	*Osphronemus olfax*
Grey Seal	*Halichoerus grypus*
Grey Squirrel	*Sciurius carolinensis*
Gulls	Laridae family
Herring Gull	*Larus argentatus*
House Sparrow	*Passer domesticus*
Kite (Red)	*Milvus milvus*
Manatee	*Trichechus manatus*
Mouse (Harvest)	*Mus minutus*
Mouse (House)	*Mus musculus*
Mink	*Mustela lutreola*
Mongoose	Viverridae family
Montagu's Harrier	*Circus pygargus*
Mina Bird (Common or Indian)	*Acridotheres cristatellus*
Oysters	Ostreidae family
Peacock	*Pavo cristatus*
Peregrine Falcon	*Falco peregrinus*
Pheasant	*Phasianus colchicus*
Rabbit	*Oryctolagus cuniculatus*
Red Squirrel	*Sciurius vulgaris*
Skua (Great)	*Catharacta skua*
Sparrowhawk	*Accipiter nisus*
Starling	*Sturnus vulgaris*
Sunfish	*Filapia mossambica*
Weddell Seals	*Leptonychotes weddelli*
Wood-pigeon	*Columba palumbus*

Appendix 3

PESTICIDES

I. INSECTICIDES

Common Name	Chemical Name
'Abate' (trade name)	
Aldicarb	2-methyl-2-(methylthio)-propionaldehyde o-(methylcarbamoyl) oxime.
Aldrin	1,2,3,4,10-hexachloro-1,4,4a,5,8,8a-hexahydro-*exo*-1,4-*endo*-5,8-dimethano-naphthalene
Aminocarb	4-dimethylamino-3-methylphenyl N-methylcarbamate
Aphoxide	*tris*-(1-azaridimyl)phosphine oxide
Asulam	methyl (4-aminobenzenesulphonyl) carbamate
Azinphos-methyl	S-(3,4-dihydro-4-oxobenzo[*d*]-[1,2,3]-triazin-3-ylmethyl) dimethyl phosphorothiolothionate
Azobenzene	diphenyl diimide
BHC	mixed isomers of 1,2,3,4,5,6-hexachlorocyclohexane
Carbon disulphide	carbon disulphide
Carbaryl	1-naphthyl methylcarbamate
Chlorbenzide	4-chlorobenzyl 4-chlorophenyl sulphide
Chlordane	1,2,4,5,6,7,10,10-octachloro-4,7,8,9-tetrahydro-4,7-methyleneindane
Chlorfenethol	1,1-bis-(4-chlorphenyl)ethanol

194

Common Name	*Chemical Name*
Chlorfenson	4-chlorophenyl-4-chlorobenzene sulphonate
Chlorodibromopropane	chlorodibromopropane
Chloropicrin	trichloronitromethane
Chlorpyrifos	diethyl 3,5,6-trichloropyridyl phosphorothioate
Copper sulphate	copper disulphate
Coumaphos	3-chloro-4-methyl-7-coumarinyl diethyl phosphorothionate
Creosote	creosote
Cresylic acid	cresylic acid
Crotoxyphos	dimethyl-2-(α-methylbenzoxycarbonyl)-1-methyl vinyl phosphate
Crufomate	4-butyl-2-chlorophenylmethyl-N-methyl phosphoroamidate
DCPM	bis(4-chlorophenoxy)methane
DD	mixture of 1,3-dichloropropene, 1,2-dichloropropane, and 2,3-dichloropropene
DDT	1,1,1-trichloro-2,2-di-(4-chlorophenyl) ethane
DDD (or TDE)	1,1,dichloro-2,2-di-(4-chlorophenyl) ethylene
Demeton	Mixture of diethyl 2-(ethylthio)ethyl phosphorothionate (demeton o) and diethyl 2-(ethylthio)ethyl phosphorothiolate
Demeton-S-methyl	S-(ethylthio)ethyl dimethyl phosphorothiolate
Diazinon	oo-diethyl o-(2-iso-propyl-6-methyl-4-pyrimidinyl) phosphorothioate
Dichlorocresylic acid	dichlorocresylic acid
Dicofol	2,2,2-trichloro-1,1-di-(4-chlorophenyl) ethanol
Dicrotophos	dimethyl-2-dimethylcarbamoyl-1-methylvinyl phosphate (E isomer)
Dieldrin	1,2,3,4,10,10-hexachloro-6,7-epoxy-1,4,4a,5,6,7,8,8a-octahydro-*exo*-1,4-*endo*-5,8-dimethano-naphthalene
Dimefox	NNN′N′-tetramethylphosphorodiamidic fluoride
Dimetilan	2-dimethylcarbamoyl-3-methyl-5-

Appendices

Common Name	*Chemical Name*
	pyrazolyl dimethylcarbamate
Dimethoate	oo-dimethyl S-(N-methylcarbamoylmethyl) phosphorothiolothionate
DNOC	3,5-dinitro-o-cresol
Endosulfan	6,7,8,9,10,10-hexachloro-1,5,5a,6,9,9a-hexahydro-6,9-methano-2,4,3,-benzoldioxathiepin-3-oxide
Endrin	1,2,3,4,10,10-hexachloro-6,7-epoxy-1,4,4a,5,6,7,8,8a-octahydro-*exo*-1,4-*exo*-5,8-dimethanonaphthalene
Ethylene dibromide	1,2-dibromoethane
Ethylene oxide	ethylene oxide
'Eulan' (trade name)	
Fenitrithion	dimethyl 3-methyl-4-nitrophenyl phosphorothionate
Fenchlorphos	dimethyl 2,4,5-trichlorophenyl phosphorothionate
Fenthion	dimethyl 3-methyl-4-methylthiophenyl phosphorothionate
Gamma-BHC	gamma isomer of 1,2,3,4,5,6-hexachlorocyclohexane
Heptachlor	1,4,5,6,7,8,8-heptachloro-3a,4,7,7a-9-tetrahydro-4,7,-methanoindene
Hydrogen cyanide	hydrogen cyanide
'Landrin' (trade name)	
'Lethane' (trade name)	
Lindane	see gamma-BHC
Malathion	S-[1,2-di(ethoxycarbonyl)ethyl] dimethyl phosphorothiolothionate
Menazon	S-(4,6-diamino-1,3,5-triazin-2-ylmethyl) phosphorothiolothionate
Metaldehyde	metaldehyde
Metham sodium	sodium N-methyldithiocarbamate
Methiocarb	4-methylthio-3,5-xylyl N-methylcarbamate
Methoxychlor	1,1,-trichloro-2,2-di(4-methoxyphenyl) ethane
Methyl bromide	methyl bromide
Mevinphos	2-methoxycarbonyl-1-methylvinyl dimethyl phosphate
Naphthalene	naphthalene

196

Common Name	*Chemical Name*
Niclosamide	5-chloro-N-(2-chloro-4-nitrophenyl) salicylamide
Nicotine	3-(1 methyl-2-pyrrolidinyl)pyridine
Paradichlorobenzene	paradichlorobenzene
Parathion	diethyl 4-nitrophenyl phosphorothionate
Paris green	copper aceto-arsenite
Pentachlorophenol	pentachlorophenol
Phosphamidon	2-chloro-2-diethylcarbamoyl-1-methylvinyl dimethyl phosphate
Propoxur	2-isopropoxyphenyl N-methylcarbamate
Pyrethrins	mixed esters of pyrethrolone and cinerolone with chrysanthemic and pyrethric acids
Schradan	Bis-NNN'N'-tetramethylphosphorodiamidic anhydride
TEPP	tetraethylpyrophosphate
Tetradifon	2,4,4,5-tetrachorodiphenyl sulphone
Trichlorphon	dimethyl 2,2,2,-trichloro 1-hydroxyethyl phosphonate
Tricresylphosphate	tricresylphosphate
Trifenomorph	N-trityl morpholine
'Zectran' (trade name)	

2. HERBICIDES

Aminotriazole (ATA)	3-amino-1,2,4-triazole
Ammonium sulphamate	ammonium sulphamate
Asulam	methyl(4-aminobenzenesulphonyl) carbamate
Atrazine	2-chloro-4-ethylamino-6-isopropylamino-1,3,5-triazine
Aziprotryne	2-azido-4-isopropylamino-6-methylthio-1,3,5 triazine
Barban	4-chlorobut-2-ynyl-N-(3-chlorophenyl) carbamate

o

Common Name	*Chemical Name*
Benzoylprop-ethyl	ethyl 2-(N-benzoyl-3,4-dichloroanilino) propionate
Bromacil	5-bromo-6-methyl-3-(1-methylpropyl) uracil
Bromoxynil	3,5-dibromo-4-hydroxybenzonitrile
Chlorflurecol methyl	2-chloro-9-hydroxyfluorene -9-carboxylic acid
Chlormequat	2-chloroethyltrimethylammonium ion
Chloroxuron	N^1-4-(4-chlorophenoxy) phenyl-NN-dimethylurea
Chlorpropham	isopropyl N-(3-chlorophenyl) carbamate
Chlorthiamid	2,6-dichlorothiobenzamide
Chlortoluron	N'-(3-chloro-4-methylphenyl)-NN-dimethyl urea
Dalapon	2,2-dichloropropionic acid
Daminozide	N-dimethylamin; succinamic acid
Desmetryne	2-isopropyl amino-4-methyl amino-6-methyl thio-1,3,5 triazine
Di-allate	S-2,3-dichloroallyl NN-diisopropylthiolcarbamate
Dicamba	3,6-dichloro-2-methoxybenzoic acid
Dichlobenil	2,6-dichlorobenzonitrile
2,4-D	2,4-dichlorophenoxyacetic acid
2,4-DB	4-(2,4-dichlorophenoxy) butyric acid
Dichloroprop	($\pm$)2-(2,4-dichlorophenoxy) propionic acid
Dimexan	di(methoxythiocarbonyl) disulphide
Dinatramine	N^3N^3-diethyl 2,4-dinitro-6-trifluoromethyl-m-phenylenediamine
DNOC	3,5-dinitro-o-cresol
Dinoseb	2-(1-methyl-n-propyl)-4,6-dinitrophenol
Diquat	9,10-dihydro-8a,10a-diozoniaphenanthrene ion
Diuron	3-(3,4-dichlorophenyl)-1,1-dimethylurea
Fenoprop	($\pm$)2-(2,4,5-trichlorophenoxy)propionic acid
Fenuron	1,1-dimethyl-3-phenylurea
Glyphosphate	N-(phosphonomethyl) glycine
Ioxynil	4-hydroxy-3,5-diiodobenzonitrile
Indole-3-acetic acid	indole-3-acetic acid

Common Name	Chemical Name
Lenacil	3-cyclohexyl-6,7-dihydro-1H-cyclopentapyrimidine-2,4-(3H, 5H)-dione
Linuron	3-(3,4-dichlorophenyl)-1-methoxy-1-methylurea
Maleic hydrazide	6-hydroxy-3-(2H)-pyridazinone
MCPA	4-chloro-2-methylphenoxyacetic acid
MCPB	4-(4-chloro-2-methylphenoxy) butyric acid
Mecoprop	($\pm$)2-(4-chloro-2-methylphenoxy) propionic acid
Methazole	2-($3^1,4^1$-dichlorophenyl)-4-methyl-1,2,4-oxodiazolidine-3,5 dione
Metribuzon	4 amino-3-methylthio-6-t-butyl-1,2,4- triazin-5(4 H)-one
Monlinuron	3-(4-chlorphenyl)-1-methoxy-1-methylurea
Monuron	3-(4-chlorophenyl)-1,1-dimethylurea
Neburon	1-n-butyl-3-(3,4-dichlorophenyl)-1-methylurea
Paraquat	1,1^1-dimethyl-4,4^1-bipyridylium
PCP	pentachlorophenol
Prometryne	2,4-bis(isopropyl amino)-6-methyl thio-1,3,5 triazine
Propachlor	2-chloro-N-isopropyl acetanilide
Propanil	N-(3,4-dichlorophenyl)propionamide
Propazine	2-chloro 4,6-bisisopropylamino-1,3,5-triazine
Simazine	2-chloro-4,6-bis(ethylamino)-1,3,5-triazine
Sodium chlorate	sodium chlorate
Terbacil	3-t-butyl-5-chloro-6-methyluracil
Tri-allate	S-2,3,3-trichloroallyl diisopropylthiolcarbamate
Trifluarin	2,6-dinitro-NN-dipropyl-4-trifluoromethylaniline
TCA	trichloroacetic acid
2,3,6-TBA	2,3,6-trichlorobenzoic acid
2,4,5-T	2,4,5-trichlorophenoxyacetic acid

Common Name	*Chemical Name*

3. FUNGICIDES

'Agrosan G' (trade name)

Benomyl	methyl 1-butylcarbamoyl-2-benzimidazol-2-ylcarbamate
Binapacryl	2-(1-methyl-n-propyl)-4, 6-dinitrophenyl-2-methylcrotonate
Biphenyl	
Captafol	N-(1,1,2,2-tetrachloroethylthio)-3a,4, 7, 7a-tetrahydrophthalimide
Captan	N-(trichloromethylthio)3a,4,7,7a-tetrahydrophthaliamide
Carboxin	2,3-dihydro-6-methyl-5-phenylcarbamoyl-1,4-oxathiin
'Ceresan' (trade name)	
Chloranil	2,3,5,6-tetrachloro-1,4-benzoquinone
Dichlone	2,3-dichloro-1,4-naphthoquinone
Dimethirimol	5-n-butyl-2-dimethylamino-4-hydroxy-6-methylpyrimidine
Dinocap	2-(1-methyl-n-heptyl)-4,6-dinitrophenyl crotonate
Fentin acetate	triphenyltin acetate
Fentin hydroxide	triphenyltin hydroxide
Folpet	N-(trichloromethylthio)phthalimide
'Germison' (trade name)	
Maneb	manganese ethylene 1,2-bisdithio carbamate
Metham	N-methyldithiocarbamic acid
'Panogen' (trade name)	
Thiram	bis(dimethylthiocarbamoyl) disulphide
Triamiphos	5-amino-1-bis(dimethylamido) phosphinyl-3-phenyl-1,2,4-triazole
'Upsulin' (trade name)	
Zineb	zinc ethylene-1,2-bisdithiocarbamate

4. PESTICIDES USED AGAINST VERTEBRATES

Norbomide	5-(α-hydroxy-α-2-pyridylbenzyl)-7-(α-2-pyridylbenzylidene)norbor-5-ene-2,3-dicarboxyimide
Warfarin	3-(α-acetonylbenzyl)-4-hydroxycoumarin

Index

201

P

Index

Index

Index

Index

Index

In his lengthy struggle for survival, Man has had to fight a constant campaign against the pests that threaten him. In past ages, his arms were few and pitifully inadequate, so that both his numbers and his standard of living were severely restricted. Now, with the advent of pesticides and other novel methods of control, Man is for the first time gaining the upper hand. The struggle is, however, a continuing one, and there can be no grounds for complacency: the enemy, too, holds some powerful weapons, not least the development of resistant strains.

Pesticides are described and discussed; but these are not the only means used to combat pests. Biological control—that is, the utilization of pests' natural enemies against them—is also examined, as are other new techniques including the employment of sex-attractants and the mass sterilization of male insects.

This book provides a comprehensive study of all types of pest control, together with a balanced view of the advantages and disadvantages of pesticide usage. While it is an authoritative survey, *The Pest War* will also be of interest to the non-technical reader.

The author is Professor of Biology at the University of Strathclyde, and is internationally known as an expert on pesticides.